FOUNDATIONS OF
BOUNDARY
LAYER THEORY
for Momentum, Heat, and Mass Transfer

Joseph A. Schetz

Aerospace and Ocean Engineering Department
Virginia Polytechnic Institute and State University

FOUNDATIONS OF BOUNDARY LAYER THEORY

for Momentum, Heat, and Mass Transfer

PRENTICE-HALL, INC.
Englewood Cliffs, New Jersey 07632

Library of Congress Cataloging in Publication Data

SCHETZ, JOSEPH A.
  Foundations of boundary layer theory for momentum,
heat, and mass transfer.

  Bibliography: p.
  Includes index.
  1. Boundary layer. 2. Momentum transfer. 3. Heat—
Transmission. 4. Mass transfer. I. Title.
QA913.S35 1984   532′.05   83-7318
ISBN 0-13-329334-3

© 1984 PRENTICE-HALL, INC.
Englewood Cliffs, N.J. 07632

Printed in the United States of America

10 9 8 7 6 5 4 3 2 1

ISBN: 0-13-329334-3

Editorial/production supervision
and interior design by Tom Aloisi
Cover design by Ray Lundgren
Manufacturing buyer: Tony Caruso

Prentice-Hall International, Inc., *London*
Prentice-Hall of Australia Pty. Limited, *Sydney*
Prentice-Hall Canada Inc., *Toronto*
Editora Prentice-Hall do Brazil, Ltda., *Rio de Janeiro*
Prentice-Hall of India Private Limited, *New Delhi*
Prentice-Hall of Japan, Inc., *Tokyo*
Prentice-Hall of Southeast Asia Pte. Ltd., *Singapore*
Whitehall Books Limited, *Wellington, New Zealand*

This book is dedicated to the memory of William H. Webb, builder of clipper ships and founder and endower of the Webb Institute of Naval Architecture, whose generosity made it possible for this grandson of uneducated Albanian, German, and Slovak immigrants to obtain an excellent technical education completely free.

CONTENTS

7   WALL-BOUNDED, INCOMPRESSIBLE TURBULENT FLOWS   139

8   FREE TURBULENT SHEAR FLOWS—JETS AND WAKES   220

PREFACE

This volume was planned and written as a textbook for advanced undergraduate or beginning graduate students with three primary goals. First, it was intended to present understandable coverage at that academic level of the advances in turbulence modeling and the application of large digital computers to boundary layer problems, which have revolutionized the field over the last two decades. None of the existing texts serve that purpose. Some have held that modern numerical methods cannot be taught to engineering undergraduates so that they finish the course with any usable *tools*. That view is rejected here. Indeed, this writer strongly believes that the modern university student grew up in the *computer age*, and that he or she finds this type of material easier to grasp than such classical topics as Laplace transforms or Fourier series. Repeated experience in the classroom has proven this view to be correct. The second goal was to treat mass transfer in an integrated manner with momentum and heat transfer. The phenomena and methods of analysis are so similar that it seems inefficient and confusing to split convective mass transfer off as a separate subject as has often been the practice. Finally, a determined effort has been made throughout to relate viscous phenomena in general to the real world.

    This book is written to be applicable for courses for mechanical, aerospace, chemical, civil, and ocean engineering students. The treatment presumes that the student has had at least one undergraduate course in fluid mechanics. Tables following this preface suggest coverage for a one-semester or two-quarter course for different majors at both the advanced undergraduate and beginning graduate levels.

To achieve the goals set for coverage and length, it was necessary to omit any discussions of unsteady flows or truly three-dimensional cases. Those topics are, however, usually not discussed in any detail in courses at the intended level. Further, to leave room in the book and time in the classroom for thorough treatments of numerical methods and turbulent flows, much of the older material on laminar flows was also omitted. Although some of that material is quite elegant and interesting, it really has little actual use to the practicing engineer.

There is one other somewhat unusual feature to the organization of the material. Integral methods are introduced very early, before the derivation of the differential equations of motion. The purpose here was to provide the student with some tools, so that simple problems can be worked early in the course. The author has found this to be a helpful motivating factor for the student.

A book is a personal thing to any author and it obviously reflects his or her individual background, experience, and current view of the subject. This writer has been fortunate to have had the opportunity to interact with some of the most prolific workers in the field: Robert M. Drake, Jr., George Mellor, Antonio Ferri, Paul Libby, and Edward R. Van Driest. To them, sincere thanks are due. Special thanks are due Roger Eichhorn, who taught me the value of a combined experimental and analytical approach to any new boundary layer problem. Finally, several people were kind enough to read early versions of the manuscript and provide constructive comments. My thanks to David Rooney, Heehwan Lee, George Wills, Dinshaw Contractor, Herman Krier, Felix Pierce, and George Inger.

Joseph A. Schetz
Blacksburg, Va.
July 1982

Upper Division Mechanical or Chemical Engineering	Upper Division Aerospace Engineering	Upper Division Civil or Ocean Engineering
Chap. 1	Chap. 1	Chap. 1
Chap. 2	Chap. 2	Chap. 2 (Skip Secs. 2-4, 2-5, 2-6, and 2-7)
Chap. 3	Chap. 3	Chap. 3 (Skip Secs. 3-4 and 3-5)
Chap. 4	Chap. 4	Chap. 4 (Skip Secs. 4-4 and 4-5)
Skip Chap. 5	Chap. 5 (Skip Secs. 5-4, 5-6, and 5-8-1)	Skip Chap. 5
Chap. 6 (Skip Secs. 6-2, 6-3, and 6-5-6)	Chap. 6 (Skip Secs. 6-2 and 6-3)	Chap. 6 (Skip Secs. 6-2, 6-3, and 6-5-6)
Chap. 7 (Skip Secs. 7-10, 7-11, 7-12, 7-13, and 7-14)	Chap. 7 (Skip Secs. 7-10, 7-11, 7-12, 7-13, and 7-14)	Chap. 7 (Skip material on injection and Secs. 7-10, 7-11, 7-12, 7-13, and 7-14)
Chap. 8 (Skip Secs. 8-4, 8-5, 8-6, and 8-7)	Skip Chap. 8	Chap. 8 (Skip Secs. 8-2-2, 8-2-3, variable-density material, 8-3-2, 8-4, 8-5, 8-6, and 8-7)
Chap. 9 (Skip high-speed-flow material in Secs. 9-2-1 and 9-2-2, and Secs. 9-3-3, 9-3-4, and 9-3-5)	Chap. 9 (Skip Secs. 9-3-3, 9-3-4, and 9-3-5)	Skip Chap. 9
		Ocean Engineering
		Add outside material on free surface effects
		Civil Engineering
		Add outside material on open channel flows and sedimentation

First-Year Graduate Mechanical or Chemical Engineering	First-Year Graduate Aerospace Engineering	First-Year Graduate Civil or Ocean Engineering
Chap. 1 (Review only)	Chap. 1 (Review only)	Chap. 1 (Review only)
Appendix B	Appendix B	Appendix B
Skip Chap. 2 (*or* Review only)	Skip Chap. 2 (*or* Review only)	Skip Chap. 2 (*or* Review only)
Chap. 3 (Review only)	Chap. 3 (Review only)	Chap. 3 (Review only) (Skip Secs. 3-4 and 3-5)
Chap. 4	Chap. 4	Chap. 4
Skip Chap. 5	Chap. 5	Skip Chap. 5
Chap. 6 (Skip Sec. 6-5-6)	Chap. 6	Chap. 6 (Skip Sec. 6-5-6)
Chap. 7 (Review only Secs. 7-3-1, 7-3-2, and 7-7)	Chap. 7 (Review only Secs. 7-3-1, 7-3-2, and 7-7)	Chap. 7 (Skip material on injection, review only Secs. 7-3-1, 7-3-2, 7-7)
Chap. 8	Chap. 8	Chap. 8 (Skip Secs. 8-2-2 and 8-2-3 and variable-density material in Sec. 8-3-2)
Chap. 9 (Skip high-speed-flow material)	Chap. 9	Skip Chap. 9
		Civil Engineering *Add* advanced outside material on open channel flow and sedimentation
		Ocean Engineering *Add* advanced outside material on free surface effects

NOTATION

a	Speed of sound and amplification factor
A	Area or constant
$b_{1/2}$	Half-width
B_1, B_2, B'_1, B'_2	Constants
c	Average speed of molecules
c'_i	Fluctuating value of species concentration
c_i	Species concentration
C_i	Mean value of species concentration
C_1, C_2, etc.	Constants
c_p	Specific heat at constant pressure
c_v	Specific heat at constant volume
c_r, c_i	Real and imaginary parts of the phase velocity
C_f	Skin friction coefficient
$\bar{C}_f$	Average skin friction coefficient
C_p	Pressure coefficient
C_D	Drag coefficient
D, d	Diameter
D_{ij}	Binary diffusion coefficient
D_T	Turbulent diffusion coefficient
e	Internal energy
$f(\cdot)$	Function of $(\cdot)$

f_i	Body force vector
g	Acceleration of gravity
h	Enthalpy
$\hbar$	Film coefficient
$\hbar_D$	Film coefficient for diffusion
$H(\Lambda)$	Shape factor
i	$\equiv \sqrt{-1}$
j	Index
J	Integrated momentum flux
k	Thermal conductivity and average roughness size
k_T	Turbulent thermal conductivity
k_1	Wave number of fluctuations
K_c, K_{cp}	Mass transfer parameters
K_1, K_2, etc.	Constants
K	Turbulent kinetic energy
$E_1(k_1)$	Kinetic energy of axial fluctuations at wave number k_1
ℓ	Turbulent scale length
ℓ_m	Mixing length
Le	Lewis number
Le_T	Turbulent Lewis number
m	Index along surface
$\dot{m}_i$	Diffusive mass flux of species i
M	Maximum value of m and Mach number
n	Index across layer
N	Maximum value of n
Nu	Nusselt number
Nu_{Diff}	Nusselt number for diffusion
p	Pressure
p_i	Partial pressure of species i
P	Mean pressure
P_C, P_T, P_V	Power law decay exponents
p'	Fluctuating pressure
Pr	Prandtl number
Pr_T	Turbulent Prandtl number
$\mathscr{P}$	Production of turbulent kinetic energy
q_i	Heat flux vector
q_T	Turbulent heat flux
q_w	Wall heat transfer rate

r	Radial coordinate and recovery factor
R	Pipe radius, gas constant, and radius of curvature
Ri	Richardson number
$r_0(x)$	Body radius
$r_{1/2}$	Half-radius
Re	Reynolds number
s	Transformed streamwise coordinate
Sc	Schmidt number
Sc_T	Turbulent Schmidt number
St	Stanton number
St_{Diff}	Stanton number for diffusion
$S(\Lambda)$	Shear parameter
t	Time
T	Static temperature
$\mathbf{T}_{x, y, z}$	Surface force vector
T_b	Bulk temperature
T^*	Reference temperature
T_t	Total (stagnation) temperature
T_0	Time period
$\bar{T}$	Mean temperature
T'	Fluctuating temperature
T_*	Heat transfer temperature
T^+	$\equiv (T_w - \bar{T})/T_*$
u	Streamwise velocity
u_{ave}	Average velocity
U	Mean velocity
$\tilde{U}$	Mass-weighted mean velocity
u'	Fluctuating velocity
u_*	Friction velocity
u^+	$\equiv U/u_*$
v	Transverse or radial velocity
v_w	Transverse velocity at the wall
V	Mean transverse velocity and general velocity
v_0^+	Dimensionless transverse velocity at the wall
v'	Fluctuating transverse velocity
x	Streamwise coordinate
X_i	Mole fraction of species i
y	Transverse coordinate

Y	Transformed transverse coordinate
$y^+ \equiv yu_*/v$	Transverse coordinate for the law of the wall
W_i	Molecular weight
$W(y/\delta)$	Wake function
Z	$\equiv k^m l^n$

Greek

α	Wave number and amplification factor
α_T	$\equiv k_T/\rho c_p$
β	Pressure gradient parameter and wave number
ψ	Planar stream function
Ψ	Axisymmetric stream function
$\hat{\psi}$	Disturbance stream function
ε	Dissipation of turbulent energy
$\varepsilon_{n,m}$	Truncation error
ε_{xy}	Strain
ρ	Density
λ	Pohlhausen pressure gradient parameter, pipe resistance coefficient, and second viscosity coefficient
λ^*	Mean free path between molecules
Λ	Thwaites–Walz pressure gradient parameter
Λ_t	Smith and Spalding parameter
τ, τ_{xy}	Shear
τ_T	Turbulent shear
Ω	Intermittency
μ	Laminar viscosity
μ_T	Turbulent viscosity
κ	Constant in the law of the wall
κ_T	Constant in the Temperature law of the wall
v	Laminar kinematic viscosity
v_T	Turbulent kinematic viscosity
ϕ	Amplitude function
δ	Boundary layer thickness
δ_t	Conduction thickness
δ_T	Thermal boundary layer thickness
δ_c	Concentration boundary layer thickness
δ^*, Δ^*	Displacement thickness
δ_k^*	Kinematic displacement thickness

θ	Momentum thickness
Θ, θ_r	Excess temperature
Δ	Clauser integral boundary layer thickness
ζ	$\equiv \delta_T / \delta$
$\phi_{1,2}$	Deformation angle
Π	Wake parameter
ξ	Dummy variable
η	Similarity variable
$\bar{\eta}$	Transformed transverse coordinate
ω	Dimensionless frequency and viscosity law exponent
$\sigma_{K,z,\tau}$	"Prandtl" numbers for K, Z, τ
Γ	See Eq. (9-8)
γ	Ratio of specific heats

Subscripts

c	Values on the centerline
e	Values at the edge of the boundary layer
j	Initial values in a jet
t	Stagnation values
w	Wall values
∞	Conditions in the approach flow

**FOUNDATIONS OF
BOUNDARY
LAYER THEORY**
for Momentum, Heat, and Mass Transfer

INTRODUCTION
TO VISCOUS FLOWS

1-1 THE IMPORTANCE OF VISCOUS PHENOMENA

An applied fluid dynamicist is generally engaged in the analysis and/or design of a physical device which has a specific practical purpose (e.g., an airplane, a piping system, or a heat exchanger). In virtually all such situations, the cost of operation as well as performance is of primary importance, and this usually comes down to estimating the resistance to fluid motion. For a vehicle (e.g., a submarine), this is called *drag*. For a piping system, one must determine the *pressure loss*. In other cases, we may speak simply of *frictional losses* or in rotating electrical machines, *windage*. In general, it should be clear, then, that the influence of fluid friction can seldom, if ever, be disregarded in actual engineering practice. Thus, the *inviscid fluid* assumption is indeed too restrictive for most real situations.

A process that is closely allied to fluid friction is convection heat transfer. This subject is very important in cases ranging from the heat exchanger in an industrial boiler to a reentry vehicle. The close linkage that exists between fluid friction and convection phenomena both on basic physical grounds and on their importance from a practical point of view makes the simultaneous study of these processes attractive, and that path will be followed in this book.

A third process related to viscous resistance and convection heat transfer is mass transfer in the presence of fluid motion. Practical examples include condensation on a cool surface, industrial drying operations, and fuel droplet evaporation. Thus, it has been decided here to pursue this process simultaneously with viscous effects and convection heat transfer.

It has been stressed that the subject matter of interest is concerned with real phenomena. The reader will find it helpful, therefore, to employ his or her powers of observation and recollection while studying this material. Take every opportunity to note and learn from the fluid flow examples that constantly surround you in the real world. Observe the cloud patterns, *con* trails from airplanes, and smoke plumes from industrial stacks in the sky. Watch the water flow in your bathtub or shower or behind your boat. You can easily find a myriad of other examples yourself. Try to relate them to the material under study. If you cannot, try to find a situation in your world that involves the subject of interest at the moment. This may all sound too philosophical, but the student who tries to follow this advice will find it very helpful in gaining real understanding.

1-2 CONDITIONS AT A FLUID/SOLID BOUNDARY

The mathematical basis of modern fluid dynamics was developed mostly before the true nature of matter was well understood. Thus, the fluid was considered a homogeneous, continuous medium rather than consisting of rather widely spaced molecules as we now know is the case for gases at modest pressure. With very few exceptions, such as the highly rarefied air at the outer edge of the atmosphere, it has been possible and convenient to retain that mathematical formulation and apply it successfully to the real world by making certain assumptions. For our purposes here, the most important of these concern the conditions at the interface between the fluid and a solid surface. On the scale of an air molecule, even a smoothly machined surface looks like rough terrain. Each molecule striking the surface will have numerous collisions between the various peaks and valleys and other molecules and will exchange some of its momentum and thermal energy with the surface. It is clearly an awesome task to try to follow each molecule through its history and then sum up on a statistical basis to find the average momentum and thermal energy exchange at a given location on the surface. Experiments and analytical estimates have shown that an entirely adequate representation of physical reality is contained in the assumptions that the fluid at the surface loses all its momentum relative to the surface in the interaction and that the fluid at the surface comes to the same temperature as the surface. An essentially equivalent assumption is made for the case of mass transfer, but the matter can become somewhat more complicated. A detailed discussion is reserved until mass transfer is treated in depth in a later chapter.

The result of the assumptions stated above is velocity and temperature profiles in the region near a solid surface resembling those shown in Fig. 1-1. Denoting the local velocity in the streamwise direction as u and the distance normal to the surface as y, the profile $u(y)$ at a given value of the streamwise coordinate x will look as shown. The velocity at the surface is taken as zero (this is often called the *no-slip condition*). The no-slip condition leads to the

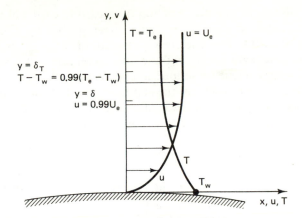

Figure 1-1 Typical velocity and temperature profiles in a boundary layer showing the definition of the velocity and temperature boundary layer thicknesses.

clarifying notion that a surface is a *sink* for fluid momentum. The velocity at a distance far (on the scale of the viscous region) from the surface will have some value dependent on the free-stream velocity U_∞ and the body shape. Call this value U_e to indicate conditions at the outer edge of the viscous region. Between the values of 0 and U_e, we expect the velocity profile to change smoothly due to the influence of fluid friction on the fluid in the layers just above and below any point of interest. The profile will blend into the edge value U_e only asymptotically, so it is common to identify the point where 99% of U_e is reached as the thickness δ of the viscous layer on the solid boundary (call this the *boundary layer*).

In a similar way, the temperature profile will vary from $T = T_w$ at the wall to $T = T_e$ at the edge of the *thermal boundary layer* δ_T. We define δ_T as the point where $T - T_w = 0.99(T_e - T_w)$. Note that δ_T is not necessarily equal to δ; that is, the thermal boundary layer thickness does not have to be the same as the velocity boundary layer thickness.

For flows with mass transfer, a composition profile across the layer will resemble those for velocity and temperature just discussed. There will be a *concentration boundary layer* of thickness δ_c.

1-3 LAMINAR TRANSPORT PROCESSES

The simplest class of flows where viscous phenomena are important occurs when the streamlines form an orderly, roughly parallel pattern. The fluid in the viscous region may be thought of as proceeding along in a series of layers or laminates with smoothly varying velocity and temperature from laminate to laminate. Consider a deck of playing cards resting on a table, and push the top several cards slightly to one side. Friction from card to card will cause most of the rest of the cards to shift in the same direction, but by a lesser amount down through the deck. The displaced pattern of the edge of the deck may be imagined to be a velocity profile, and the flow so represented would then have the laminated characteristic described above. Viscous flows of this class are, therefore, called *laminar*. They are the simplest type, but unfortunately only a fraction of flows of practical interest falls into this class. Nonetheless, some flows do, and this is a good place to begin the study of the details of viscous processes.

The reader is likely to be surprised at the simplicity and crudity of the physical arguments that form the basis of the mathematical representation of laminar flows, especially in light of the complexity of the mathematical problem that results. Some perspective on this may be gained by noting a little history. The physical arguments and the formulation of the additional viscous terms, in their present form, to augment Euler's inviscid equations of motion were accomplished in the early 1800s by Navier (1823) and Stokes (1845). It was not until the mid-1900s that practical problems could be solved based on the full formulation using the Navier–Stokes equations and only then using large digital computers. Such solutions are still by no means routine. Fortunately, a very imaginative simplification was developed just after the turn of the last century that can be used in many situations of engineering interest. We shall introduce that concept in the next section.

One can develop a suitable mathematical representation for the viscous terms by assembling some physical facts. First, a *fluid* will not sustain a stress as a result of a simple, static displacement. The fluid will simply shift and then return to rest in a new stressless state. That is the important basic mechanical difference between a fluid and a solid. If stresses in a fluid are to exist, they must result from relative motion, not position. Thus, stress in a fluid can only occur in a region of velocity variation. We would also expect that a greater rate of variation would produce a greater stress. The simplest representation that meets these criteria is

$$\tau \sim \frac{\partial V}{\partial n} \qquad (1\text{-}1)$$

Here τ is viscous stress, V the velocity, and n the direction normal to the direction of the shear stress. Note that Eq. (1-1) is based on the rather bold assumption that only the first derivative is important. We might have included other terms proportional to, for example, $\partial^2 V / \partial n^2$, but we choose to try and keep the formulation as simple as possible.

The adequacy of Eq. (1-1) can be tested in the simple device shown in Fig. 1-2, called a *viscometer*. The no-slip condition at both solid surfaces and the small gap thickness produce the linear velocity profile shown. The frictional stress can be determined from the power required to turn the cylinder. It is found that many common fluids (e.g., air, water) obey the relation in Eq.

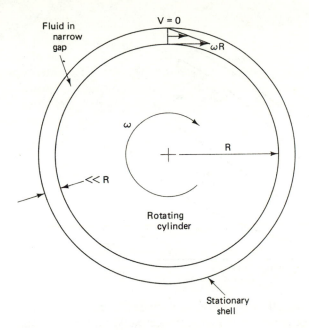

Figure 1-2   Schematic of a device for measuring fluid viscosity.

(1-1) if we write

$$\tau = \mu \frac{\partial V}{\partial n} \tag{1-2}$$

where μ is the *laminar coefficient of viscosity*, which depends strongly on the composition of the fluid. This is referred to as *Stokes' law*. Fluids that obey this law are termed *Newtonian*; others are termed *non-Newtonian*. In non-Newtonian fluids, such as polymers, the shear stress depends on the rate of shear strain, as shown in Fig. 1-3. A detailed study of the flow of such fluids is outside the scope of this book. The quantity μ is called a *physical property* of the fluid itself, since it does not depend on the state of fluid motion. Experiment shows that essentially $\mu = \mu(T)$ for a given fluid; the influence of pressure is negligible in most cases.

Looking at Eq. (1-2), we find that the dimensions of the relation suggest that

$$\mu \sim \text{density} \times \text{velocity} \times \text{length} \tag{1-3}$$

Indeed, an idealized analysis based on molecular processes indicates that for gases,

$$\mu = 0.49\rho c \lambda^* \tag{1-4}$$

where ρ is the density, c the average speed of the molecules ($c \sim \sqrt{T}$), and λ^*

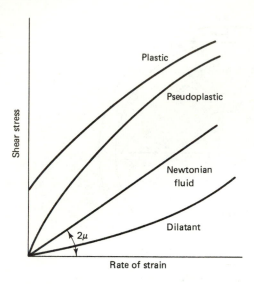

Figure 1-3 Variation of shear stress versus rate of strain for various types of fluids.

is the mean free path between the molecules. Actually, analytical predictions of μ have been of generally inadequate accuracy, and most workers use experimental data for μ when they are available. Finally, the laminar viscosity μ is often combined with the density as, $\nu \equiv \mu/\rho$, where ν is called the *laminar kinematic viscosity*.

As stated earlier, heat transfer almost always accompanies fluid friction, so we require a representation for those processes also. The physical basis consists of two facts. First, heat flow occurs only in the presence of a temperature gradient, and it is proportional to the gradient. Second, heat always flows from a higher-temperature to a lower-temperature region. Thus, we are led to

$$q = -k \frac{\partial T}{\partial n} \tag{1-5}$$

where q is the heat flux and k is the *laminar thermal conductivity*, which depends strongly on fluid composition. This is called *Fourier's law*. The minus sign ensures the correct direction of heat flow, with k always taken as positive. Experiment verifies the adequacy of Eq. (1-5) and also shows that $k = k(T)$, again with a very weak dependence on pressure. The laminar thermal conductivity is thus also a physical property of the fluid; it does not depend on fluid motion or the rate of heat transfer. Indeed, Eq. (1-5) also holds for solids.

There are two dimensionless groups involving the laminar viscosity and thermal conductivity that are useful. The first is the *Reynolds number*, named after a famous British fluid dynamicist of the late 1800s, Osborne Reynolds:

$$Re = \frac{\rho V L}{\mu} \tag{1-6}$$

Here, one must select the appropriate characteristic velocity V and length scale L for a given flow situation. By rearranging Eq. (1-6) as

$$\text{Re} = \frac{\rho V^2}{\mu V / L} \tag{1-6a}$$

we can see that this quantity is proportional to the ratio of inertia forces to viscous forces in a flow. For most flows of practical concern, this ratio is very large, on the order of 10^3 to 10^8. The second important group is the *Prandtl number*, named after a German researcher, Ludwig Prandtl, who dominated viscous flow developments in the first half of this century:

$$\text{Pr} = \frac{\mu c_p}{k} \tag{1-7}$$

Rearranging to

$$\text{Pr} = \frac{\mu / \rho}{k / (\rho c_p)} \tag{1-7a}$$

this emerges as the ratio of coefficients for diffusion of momentum to diffusion of heat. For most gases, $\text{Pr} \sim 0.7$; for liquids such as water, $\text{Pr} \sim 10$.

For the flow of a fluid consisting of a mixture of two (or more) different fluids with a nonuniform composition, there is a process of molecular diffusion akin to viscous shear and heat transfer. If the local concentration of a species is described by a *mass fraction* c_i defined as the mass of species i in a volume divided by the total mass of all fluids in that volume, the analog of Eqs. (1-2) and (1-5) for diffusion is

$$\dot{m}_i = -\rho D_{ij} \frac{\partial c_i}{\partial n} \tag{1-8}$$

This is called *Fick's law.* Here $\dot{m}_i$ is the local mass flux of species i relative to the local mass averaged velocity of the fluid. Equation (1-8) expresses the physical fact that a species diffuses from a region of high concentration to one of lower concentration. This simple expression neglects other possible processes of diffusion due to strong temperature or pressure gradients, which are often not important. The factor D_{ij} is called the *binary diffusion coefficient*, and it is also a physical property of the fluids under consideration for laminar flows. In this book, we treat in detail only mixtures of two species, that is, *binary mixtures.* The diffusion coefficient is combined with other properties to form dimensionless groupings in two ways. The first is the *Schmidt number,*

$$\text{Sc} = \frac{\mu}{\rho D_{ij}} \tag{1-9}$$

The second is the *Lewis number,*

$$\text{Le} = \frac{\rho c_p D_{ij}}{k} \tag{1-10}$$

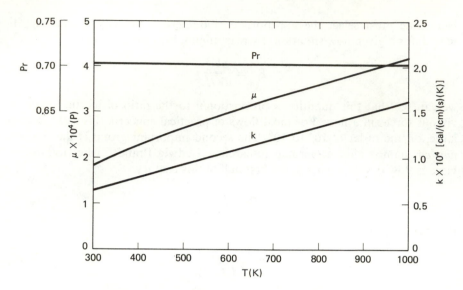

Figure 1-4   Laminar thermophysical properties for air at 1.0 atm.

Obviously they are related by

$$Pr = Le \cdot Sc \tag{1-11}$$

The Schmidt number is the ratio of coefficients for diffusion of momentum to diffusion of mass, and the Lewis number gives the ratio for mass transfer to heat.

The determination of physical properties such as the laminar viscosity, thermal conductivity, and diffusion coefficient comprises a separate field of endeavor. The work is largely composed of painstaking experiments. The working fluid dynamicist must have available ready sources of this critical information. Some data for air and water are plotted in Figs. 1-4 and 1-5. Note that the Prandtl number is nearly constant for air. That is true for most gases. Further tabulated data for a sampling of fluids are contained in Appendix A for the convenience of the reader. Appendix A also has tables of conversion factors which are helpful when data are to be obtained from a variety of sources and hence likely in a variety of systems of units.

It is also common to fit empirical information to a curve. A widely used formula for the viscosity of air is the Sutherland (1893) formula,

$$\mu = 0.1716 \left(\frac{T}{273.1} \right)^{3/2} \frac{383.7}{T + 110.6} \tag{1-12}$$

where T is in degrees Kelvin and μ is in millipoise ($N \cdot s/m^2 \times 10^4$ gives mP).

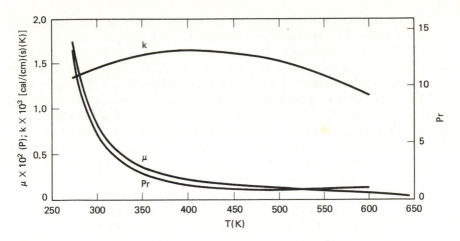

Figure 1-5 Laminar thermophysical properties for liquid water.

1-4 THE BOUNDARY LAYER CONCEPT

We have already emphasized that the Navier–Stokes equations† that result
from the complete viscous additions to the inviscid equations are very difficult
to solve for general flows. Thus, they sat largely unused for several decades,
and they did not contribute significantly to the early development of airplanes
or improved ship-hull forms, as examples. When faced with a difficult mathematical problem, it is natural to seek rigorous or at least reasonable simplifications. This is usually done by comparing the magnitude of the various
terms in the equation(s) to see if any may be neglected compared to those
retained for conditions of interest. To make this procedure rational, it is necessary to first nondimensionalize and normalize all the independent and dependent variables, preferably all to approximately order unity. For example, we
may scale the local streamwise velocity u by the free-stream velocity U_∞, and
the axial coordinate x by the length of a body, the diameter of a pipe, or other
length scale. When this is done for the Navier–Stokes equations, one is left
with the factor 1/Re, multiplying all the viscous terms. Now, as we observed
earlier, the Reynolds number is very large for most practical flows, so this
factor would be roughly 10^{-3} to 10^{-8}.

Under those circumstances it certainly seems safe to neglect all terms
multiplied by 1/Re, since the other terms in the equations have coefficients of
roughly unity, and we have sought to make all the terms themselves of order
unity. This all seems plausible; the only problem is that one is, in this way, led
back to the inviscid flow equations, which cannot predict frictional drag or
heat transfer.

† See Appendix B for a derivation of these equations.

The matter rested in this seeming paradox until the penetrating analysis of Ludwig Prandtl (1904). He reasoned that for high-Reynolds-number flows, viscous effects would be confined to a thin layer along a solid surface. Such a region must always exist, since the velocity must vary from a relatively high value out in the flow down to zero on the body surface. This will produce large gradients and hence significant shear forces even for low-viscosity (high Re) flows by virtue of Eq. (1-2). Although this region may occupy only a small portion of a total flow field, it cannot be neglected, since all momentum, heat, and mass transfer to or from the surface must take place through this layer. Outside this boundary layer, the flow does behave as if it were inviscid. Thus was born the brilliant new idea of dividing the flow into two regions and using different equations in each. For the outer region, we neglect viscous effects and use the inviscid, Euler equations. For the inner, boundary layer region, we use a simplified version of the Navier–Stokes equations that still, however, retains viscous terms. Of course, the solutions for the two regions must be suitably joined to produce a composite solution for the whole flow field. This idea is so powerful that it has been used in many other areas of science besides fluid dynamics. The term *boundary layer analysis* is, therefore, widely used in a general way.

The boundary layer equations can be derived by employing the arguments detailed above in an order-of-magnitude analysis of the Navier–Stokes equations [Eqs. (B-12)–(B-14) in Appendix B]. The single most important consequence of Prandtl's formulation is that the static pressure may be taken as constant across the boundary layer. For our uses, the following physical derivation is adequate. We presume a thin viscous layer and a body or channel shape with modest surface curvature in the flow direction. The reason for the latter restriction will become clearer in the next section. Since the layer is thin everywhere, the magnitude of the rate of growth of the layer, $d\delta/dx$, must be small. Thus, streamlines in the boundary layer must all be roughly parallel to the surface, and, by presumption, they must all have, at most, a modest curvature. This is the same as saying that the radius of curvature of the streamlines in the boundary layer R must be large. From elementary fluid mechanics, we have the relation

$$\frac{\partial p}{\partial n} = \frac{\rho V^2}{R} \qquad (1\text{-}13)$$

which says that for very large R the pressure variation normal to the streamlines (including the body surface itself) must be negligible.

The fact that the static pressure is constant across the thin boundary layer is important for two reasons. First, the Navier–Stokes equations have one equation expressing conservation of momentum for each coordinate direction. For a planar (x, y) formulation, we would have an x momentum equation and a y momentum equation. Under the boundary layer assumption, the y

momentum equation reduces to

$$\frac{\partial p}{\partial y} \approx 0 \tag{1-14}$$

This results in a considerable simplification.

The second important result of taking the static pressure constant across the boundary layer concerns the joining of the outer, inviscid solution to the inner, boundary layer solution to produce a complete, composite solution. If the static pressure is constant across the layer, the static pressure predicted by the inviscid solution at the outer edge of the boundary layer determines the static pressure all across the boundary layer. One says that the inviscid solution *impresses* the static pressure on the boundary layer. This can be carried further. From the point of view of the inviscid flow, the outer edge of the thin boundary layer is indistinguishable from the surface. Thus, an inviscid solution for the flow over the body or through the channel of interest is found first, and the predicted inviscid *pressure distribution on the surface*, p(x), is *taken as the pressure in the boundary layer*. Also, Bernoulli's equation applied to the inviscid region gives the *inviscid velocity on the surface*, which is *taken as the outer edge velocity for the boundary layer*.

$$U_e \frac{dU_e}{dx} = -\frac{1}{\rho} \frac{dp}{dx} \tag{1-15}$$

From the point of view of the boundary layer, the outer edge of the boundary layer δ is far from the surface. Therefore, the boundary condition for the velocity in the boundary layer at the outer edge of the layer is not written as $u(x, \delta) = U_e$, but rather

$$\lim_{y \to \infty} u(x, y) = U_e(x) \tag{1-16}$$

From the preceding discussion, it can be seen that the coupling between the outer, inviscid flow and the inner, boundary layer flow is all one-way. The inviscid solution for flow over the body or through the channel can be found first, independently of the boundary layer. That solution determines p(x) and $U_e(x)$ for the boundary layer solution. The boundary layer equations are then solved under those constraints.

1-5 SEPARATION AND THE KUTTA CONDITION

The information developed so far can be used to explain the important phenomenon of *boundary layer separation* and that will further lead to an explanation of the physical basis of the famous *Kutta condition*, widely used in inviscid aero/hydrodynamics. When the shape of the surface of a body immersed in a flowing fluid or of a channel with fluid flowing through it is such

(A) Visualization of laminar separation from a curved wall by Werle (1982) at ONERA. Air bubbles in water.

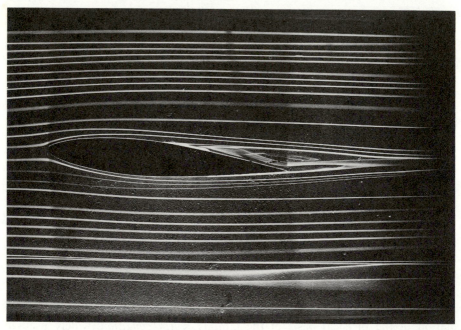

(B) Visualization of laminar separation from an airfoil at angle of attack by Werle (1982) at ONERA. Dye streaks in water.

Figure 1-6 Illustrations of the boundary layer separation process.

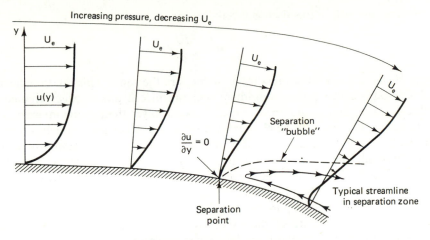

(C)   Schematic illustration of separation.

that the static pressure increases rapidly in the streamwise direction, a dramatic and undesirable change in the flow pattern is often observed. Photographs and a schematic of the flow are given in Fig. 1-6. Streamlines in the boundary layer near the surface which have been proceeding along generally following the shape of the surface suddenly depart from the surface. We say that the flow *separates* from the surface. This occurrence is undesirable because the body or channel shape was usually selected to produce a given static pressure variation for a specific purpose, such as to produce lift for a wing or to provide efficient deceleration of a flow in a diffuser. If the flow separates, the intended static pressure distribution will obviously be disrupted. Also, flow separation will disturb an intended heat or mass transfer distribution.

The processes leading to separation can be made clear by a crude but useful argument. The development is crude, not exact, because we shall use Bernoulli's equation in the viscous boundary layer, even though it strictly applies only for inviscid flow. Our purpose at this point is, however, only to illuminate the phenomena involved, not to analyze the flow in detail, so some liberties may be taken. Bernoulli's equation states that

$$u \frac{du}{ds} = -\frac{1}{\rho} \frac{dp}{ds} \tag{1-17}$$

along a streamline. In the inviscid flow outside the boundary layer, this equation will correctly relate the variation in the edge velocity, $U_e(x)$, to the pressure gradient. Note that for increasing pressure, the velocity decreases. For this exercise, we also assume that Eq. (1-17) holds locally in the boundary layer. Now, the boundary layer assumption states that the static pressure is constant across the layer, so the pressure gradient (in the streamwise direction) must also be the same in the layer. Comparing two points—one at the edge of the layer and one well in the viscous layer—we may write

$$u \frac{du}{ds} = U_e \frac{dU_e}{ds} \tag{1-18}$$

from Eq. (1-17), since the pressure gradients are the same. This relation clearly shows that a given pressure gradient will produce a much larger change in velocity for a point in the lower velocity viscous layer than in the inviscid flow at the edge. The ratio of the changes will be $U_e : u$. Since u approaches zero at the wall, a pressure gradient will produce a large Δu in a region where u is already small. The result is as shown by the successive profiles sketched in Fig. 1-6, leading finally to an actual reversal in the flow direction. This causes an effective obstacle to the upstream flow, so it *separates* from the original body surface to flow over the *separation bubble.*

When a flow separates, the viscous region is no longer thin, even at high Reynolds numbers, and the boundary layer assumptions are no longer valid. Thus, the conditions for the applicability of the boundary layer assumptions must be high-Reynolds-number flow over streamlined, not bluff, shapes.

The phenomenon of separation is important in its own right, and further discussion of it will be presented later in this book. Separation is also important in understanding the basis for the Kutta condition. The role of the Kutta condition is to select the proper, unique inviscid solution for flow over a lifting body with a sharp trailing edge, since the inviscid equations and boundary conditions provide an infinite group of solutions corresponding to various values of the *circulation.* Kutta introduced the heuristic condition that the rear stagnation point must occur at the sharp trailing edge. This amounts to inserting a crucial influence of viscosity into an inviscid solution procedure. The logic of this development can be understood with the aid of the sketches in Fig. 1-7. For a circulation lower than that set by the Kutta condition, the inviscid solution would predict a streamline pattern as shown in Fig. 1-7(A). In a real, viscous flow, the fluid near the surface coming up around the trailing edge from the underside will have lost some momentum through friction. It will then not be able to negotiate the sharp turn at the trailing edge and the adverse pressure gradient on the top surface, and it will separate. The separa-

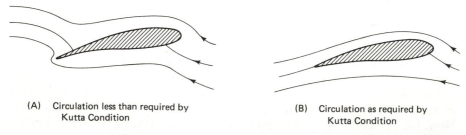

(A)   Circulation less than required by   (B)   Circulation as required by
        Kutta Condition   Kutta Condition

Figure 1-7   Streamline patterns for flow past an airfoil predicted by inviscid fluid theory.

tion point will have to be farther downstream than the inviscid stagnation point shown in Fig. 1-7(A). In fact, the only stable location will be at the trailing edge, as shown in Fig. 1-7(B). This viscous process cannot be described by the inviscid analysis, so the Kutta condition must be asserted in order to make the inviscid analysis unique. Again, we see the great importance of even small viscosity.

1-6 BASIC NOTIONS OF TURBULENT FLOW

Most flows of practical concern are not *laminar*, but rather they are *turbulent*, so it is important to gain a clear understanding of the nature of turbulent flows early in the study of viscous phenomena. In this chapter we deal only with the physical nature of turbulence. The analytical description of turbulent flows will be discussed in later chapters after a solid foundation in the form of the analytical description of laminar flows has been laid.

In Fig. 1-8 one sees beautiful visualizations of turbulent flow in the wake of a sphere from Werle (1982) at ONERA in France. It is worth the time to study these photographs carefully. Also, the reader should supplement these instantaneous examples with observations, the plume from a smokestack being a good example. The films "Characteristics of Laminar and Turbulent Flow"† and "Turbulence"‡ are both excellent and helpful. These flow visualizations reveal some of the most significant characteristics of turbulent flow. First, there is a general *swirling* nature of the flow involving indistinct lumps of fluid called *eddies*. There is a very wide range in the size of the eddies occurring at the same time or at the same place in the turbulent region. Second, the instantaneous boundary between the turbulent region and nonturbulent, outer, inviscid flow is sharp. The flow at a point near the average location of the edge of the boundary layer is intermittently either turbulent or not. Third, turbulence is basically an unsteady process. Any still picture such as those in Fig. 1-8 does not adequately convey the real situation. The unsteadiness may appear vaguely periodic, but closer study will reveal that the flow is randomly unsteady. A harmonic analysis of the motion shows that fluctuations over a very wide range of effective frequencies (several orders of magnitude) are present. Fourth, turbulence is always three-dimensional even if the background flow is two-dimensional. A final important point to be learned and remembered from flow visualizations of turbulent flows is that the irregular variations in the motion are not small with respect to either time or space. Some of the instantaneous velocity fluctuations in the wake behind a bluff body such as that shown in Fig. 1-8, for example, are of the order of magnitude of the free-stream velocity.

An instrument for measuring the fluctuating velocities in a turbulent flow

† Produced by the University of Iowa.
‡ Produced by the Encyclopaedia Brittanica.

(A) $Re_D = 15,000$. Dye streaks in water.

(B) $Re_D = 15,000$. Air bubbles in water.

(C) $Re_D = 30,000$. Air bubbles in water.

(D) $Re_D = 30,000$. Air bubbles in water.

Figure 1-8 Flow visualizations for the turbulent flow in the wake of a sphere at various Reynolds numbers from Werle (1982) at ONERA.

is the hot-wire anemometer. This is a device that is simple in concept, and modern electronics has made it easy to use. The heart of the instrument is a very thin wire (say 10^{-3} mm) stretched between two, sharp prongs that is electrically heated to a temperature slightly higher than the fluid. Fluid flowing

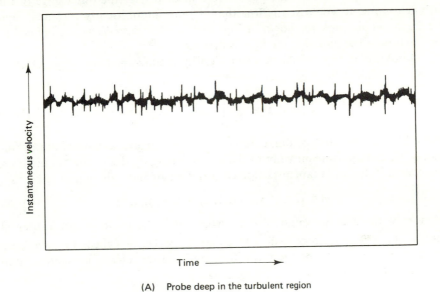

(A)   Probe deep in the turbulent region

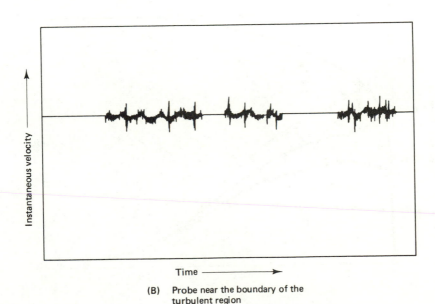

(B)   Probe near the boundary of the
turbulent region

Figure 1-9   Hot-wire anemometer signals in a turbulent flow.

past the wire will cool it by an amount depending directly on the instantaneous velocity. With very fine wires and suitable electronics, this device can follow even the very fast, small-scale components of turbulent motion. The output from a hot-wire anemometer displayed on an oscilloscope would look as shown in Fig. 1-9(A). If the probe were located near the outer edge of the turbulent zone, it would sense flow that is only intermittently turbulent, and the signal would look as shown in Fig. 1-9(B). The *intermittency* is defined as the fraction of the time that the flow is turbulent.

Looking at Fig. 1-9(A) one can identify a time-mean velocity, which is called U. This is defined as

$$U(x, y, z) \equiv \frac{1}{T_0} \int_0^{T_0} u(x, y, z, t) \, dt \qquad (1\text{-}19)$$

where T_0 is a time that is long compared to the longest periods of the fluctuations. To obtain some measure of the average magnitudes of the fluctuations, one can split the instantaneous velocity into a mean and a fluctuating part as

$$u(x, y, z, t) = U(x, y, z) + u'(x, y, z, t) \qquad (1\text{-}20)$$

Obviously, the average value of u' is zero, so we take the root-mean-squared value, $\sqrt{\overline{(u'^2)}}$. The ratio of $\sqrt{\overline{(u'^2)}}$ to either U or U_∞ is called the *turbulence intensity*. Some results for the mean flow and the turbulence intensities in all three flow directions are shown in Fig. 1-10.

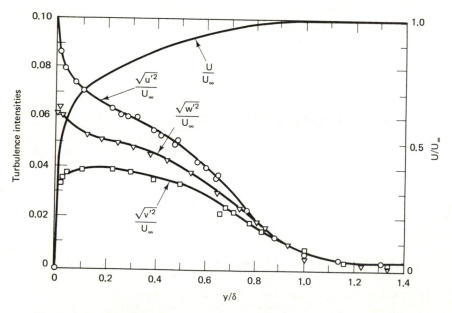

Figure 1-10 Measurements of the mean velocity and turbulent intensities in a boundary layer. (From Klebanoff, 1955).

There is a last conceptual matter that is important here. What is the difference between an unsteady, laminar flow and a turbulent flow? Following that, can we speak of a *steady*, turbulent flow? In principle, a velocity/time history like that in Fig. 1-9(A) could be produced by an unsteady, laminar flow. An example might be the flow in a laminar boundary layer near an oscillating surface. If the surface oscillations were controlled by a random number generator and they were at a high rate, the velocity near the surface could be sensed by a hot-wire anemometer, and the output might resemble that for a turbulent flow. The key to distinguishing between that case and a genuinely turbulent flow is that the velocity fluctuations in the laminar case would be directly correlated with the oscillations of the surface (i.e., the boundary conditions). The fluctuations in a turbulent flow are truly random, and they are not correlated with the boundary conditions. This leads us to the apparent contradiction in terms contained in the phase "steady, turbulent flow." The time-averaged, mean flow represented by U, for example, can indeed be steady. The mean flow behind the sphere in Fig. 1-8 will be constant with time (i.e., steady) if the cylinder is fixed and the approach flow is constant. To be precise, one says that such a flow is *steady in the mean*. A turbulent flow can also be *unsteady in the mean*. For the sphere wake case, the mean flow field would vary with time if the sphere were oscillated [i.e., $U = U(x, y, t)$]. However, any time variation of the mean component of a turbulent flow will be correlated with the time variations of the boundary conditions.

PROBLEMS

1.1. Calculate the Reynolds number based on length for a 1-m flat plate in an airstream at sea level and 25°C with a velocity of 3 m/s. Now, calculate the Reynolds number based on diameter for 25°C water flowing at 150 cm^3/s through a 1.0-cm-diameter tube.

1.2. Estimate the frictional stress in a boundary layer of flowing water 1.0 cm thick at a free-stream velocity of 5 m/s. Take the velocity gradient in the boundary layer as approximated by a linear variation of velocity from the edge conditions to the wall value. Compare this to the inertia force ρV^2. Assume ambient temperature.

1.3. Say that a hydraulic lift in an automobile shop has a shaft with a diameter of 40 cm moving in a cylinder with an inside diameter (ID) of 40.02 cm. If the shaft is moving at a speed of 0.2 m/s and the viscosity of the hydraulic fluid can be taken as that of water, what is the resistance to motion of the shaft per meter of length?

1.4. Consider an air gap of 1 mm between two flat surfaces, one at 25°C and the other at 50°C. If the air in the gap remains at rest, what is the heat transfer rate across the gap per unit area?

1.5. Locate a clearly observable turbulent flow in your surroundings. Make a series of sketches showing the instantaneous and time-varying features of that flow.

2

INTEGRAL EQUATIONS
AND SOLUTIONS
FOR LAMINAR FLOW

2-1 INTRODUCTION

Before undertaking the development of any analytical procedure, the engineer should first ask what the desired result of the analysis is to be. A practicing engineer or applied scientist is not generally interested in the *elegance* of a solution, but rather needs the answer to a specific problem. If an elegant solution provides more information than is needed at an appreciable increase in the cost of obtaining the solution compared to a cruder method, the cruder method is better on a cost/benefit basis.

What then is the output usually needed from a boundary layer analysis? Generally, the primary information required is skin friction drag, $C_f(x)$ ($\equiv \tau_w/\frac{1}{2}\rho U_e^2$). A closely related item would be the point, if any, where $\tau_w = C_f = 0$ occurs (i.e., the location of separation). If there is heating or cooling, the wall heat transfer rate, $q_w(x)$, is correspondingly sought. When there are concentration gradients, the mass transfer rate at the wall, $\dot{m}_{iw}$, is desired. The next type of information that might be useful is the transverse extent of the viscous region, $\delta(x)$ [and perhaps $\delta_T(x)$ and/or $\delta_c(x)$]. It may surprise the reader to learn that in the vast majority of practical cases, that much output is sufficient. Only rarely in a design calculation would one need to know as much detail as, for example, the streamwise velocity $u(x, y)$ at a given point in the boundary layer.

With the goal of predicting $C_f(x)$ and $\delta(x)$ [and perhaps $q_w(x)$ and $\dot{m}_{iw}(x)$], let us develop the simplest analytical method of achieving those ends. In this chapter we consider only laminar flows. There are some high-Reynolds-number cases where the Reynolds number in a substantial part of the flow is low enough for laminar flow to exist. Also, the understanding of laminar analyses is essential to attaining an understanding of the more widely applicable and more complicated turbulent analyses. To keep the derivation simple, it is helpful initially to adopt the additional restrictions of steady, incompressible (constant-density), and constant-property flow.

In formulating the appropriate analytical approach, it is important to note that all the primary desired quantities $[C_f(x), q_w(x), \dot{m}_{iw}(x), \delta(x),$ etc.$]$ are functions only of the streamwise coordinate x. Apparently, the dependence on the transverse coordinate, y, is of lesser importance, and we may be able to take some liberties. Indeed, we will treat all the flow within the boundary layer at a given station x as one unit and apply conservation principles to it as a whole.

Before doing that, perhaps it is useful to digress here a moment and say a few words about the philosophy of that approach. The engineering or physical science student generally takes this matter too lightly. By this point in his or her education, the student has seen conservation of mass, momentum, energy, and so on, applied to a variety of situations. It is taken for granted that there exist generally applicable conservation principles that can be rigorously used to produce an equation or equations that can faithfully describe real phenomena. The fact is, however, that this state of affairs exists only in this narrow branch of the spectrum of human endeavors. It does not exist in law, medicine, history, sociology, or economics, to name just a few areas. Those subjects are purely empirical; there are no basic *first principles* that have any real generality. This is why the often-heard question, "If we can send men to the moon, why can't we control our economy?" is in truth nonsensical. That is also why Isaac Newton is commonly counted in the top 10 of the most important human beings who ever lived.

2-2 THE INTEGRAL MOMENTUM EQUATION

Consider the two-dimensional flow in Fig. 2-1. The main flow is from left to right and is bounded on the bottom by a rigid surface. From the discussion in Chapter 1, it is known that the velocity at the surface $(y = 0)$ will be taken as zero [i.e., $u(x, 0) = 0$]. We expect the velocity to vary smoothly from zero at the surface and to blend into the local boundary layer edge velocity $U_e(x)$ at $y = \delta$. We shall apply our conservation principles to a control volume that is finite in the transverse direction, $y = 0$ to H [where $H > \delta(x)$], but differential in the streamwise direction dx. This is consistent with the notion of downplaying the importance of variations in the y direction.

The mass flow entering the left-hand side of the control volume is

$$\int_0^H \rho u \, dy \qquad (2\text{-}1)$$

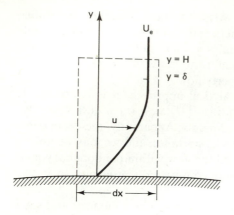

Figure 2-1 Control volume for integral momentum equation derivation.

and the mass flow leaving the right-hand side is

$$\int_0^H \rho u \, dy + \frac{d}{dx}\left[\int_0^H \rho u \, dy\right] dx \tag{2-2}$$

If the bottom is solid and the flow is steady, the mass in the control volume must remain constant, so the difference between these two must come in or go out through the top. If the wall is porous and there is flow through it v_w, then the quantity $\rho_w v_w$ must be subtracted from the difference above to obtain the flow through the top.

The momentum flow in the left-hand side is

$$\int_0^H \rho u^2 \, dy \tag{2-3}$$

and that out the right-hand side is

$$\int_0^H \rho u^2 \, dy + \frac{d}{dx}\left[\int_0^H \rho u^2 \, dy\right] dx \tag{2-4}$$

Any mass flow in the top will bring with it streamwise momentum in an amount proportional to $U_e(x)$

$$U_e(x)\left\{\frac{d}{dx}\left[\int_0^H \rho u \, dy\right] dx - \rho_w v_w \, dx\right\} \tag{2-5}$$

The net momentum flux *out* of the control volume then becomes

$$\frac{d}{dx}\left[\int_0^H \rho u^2 \, dy\right] dx - U_e(x)\left\{\frac{d}{dx}\left[\int_0^H \rho u \, dy\right] dx - \rho_w v_w \, dx\right\} \tag{2-6}$$

The change in x momentum is balanced with the summation of forces in the x direction on the fluid. Neglecting body forces such as gravity with respect to the inertia of the flow, we are left with pressure forces and shear forces. Since the boundary layer approximation is essentially $\partial p/\partial y \approx 0$ across the layer (see Chapter 1), p may be treated as $p(x)$ alone. The pressure force on the left-hand side is pH and that on the right-hand side is

$$-\left(p + \frac{dp}{dx}\,dx\right)H$$

so the net pressure force on the volume is

$$-\left(\frac{dp}{dx}\,dx\right)H \tag{2-7}$$

There is a laminar shear force on the bottom surface $(\tau_w = \mu(\partial u/\partial y)_{y=0})$ acting in the negative x direction (i.e., retarding the motion of the fluid),

$$-\tau_w\,dx = -\mu\left.\frac{\partial u}{\partial y}\right|_{y=0}\,dx \tag{2-8}$$

but none on the top surface, since $(\partial u/\partial y)_{y=\delta} = 0$. Any small shear forces on the sides of the volume have no component in the x direction.

All of this can be combined into one equation expressing conservation of momentum in the x direction. In doing so, it is convenient to rewrite the second part of Eq. (2-6) using the rule for the derivative of a product as

$$U_e(x)\frac{d}{dx}\left[\int_0^H \rho u\,dy\right]dx$$

$$= \frac{d}{dx}\left[\int_0^H \rho u\,U_e(x)\,dy\right]dx - \frac{dU_e}{dx}\left[\int_0^H \rho u\,dy\right]dx \tag{2-9}$$

Setting Eq. (2-7) plus (2-8) equal to (2-6), with (2-9), results in

$$-\tau_w - H\frac{dp}{dx}$$

$$= -\rho\frac{d}{dx}\left[\int_0^H (U_e - u)u\,dy\right] + \frac{dU_e}{dx}\,\rho\left[\int_0^H u\,dy\right] + \rho_w v_w U_e \tag{2-10}$$

This equation can be further simplified by noting that, since the pressure is assumed constant across the boundary layer, the pressure can be related to the velocity in the inviscid flow just outside the boundary layer through Bernoulli's equation,

$$-\frac{1}{\rho}\frac{dp}{dx} = U_e\frac{dU_e}{dx} \tag{2-11}$$

Finally, we introduce new notation for the integrals in Eq. (2-10) rewritten slightly:

$$\theta \equiv \int_0^\delta \left(1 - \frac{u}{U_e}\right)\frac{u}{U_e}\,dy \tag{2-12}$$

$$\delta^* \equiv \int_0^\delta \left(1 - \frac{u}{U_e}\right) dy \qquad (2\text{-}13)$$

The last integral in Eq. (2-10) was combined with the dp/dx term using Eq. (2-11) to produce (2-13). Equations (2-12) and (2-13) represent more than notational convenience, and we shall attach physical meaning to θ and δ^* shortly. In the meantime, the equation becomes

$$\frac{d\theta}{dx} + \frac{1}{U_e}\frac{dU_e}{dx}(2\theta + \delta^*) - \frac{v_w}{U_e} = \frac{C_f}{2} \qquad (2\text{-}14)$$

since $\rho = \rho_w = $ constant.

This is a good point to step back and consider what has been produced. The equation looks tractable, since it is a relatively simple, ordinary differential equation with x as the only independent variable. Our major desired output C_f appears directly as a dependent variable. Here, as with any boundary layer problem, $U_e(x)$ must be known beforehand from a prior inviscid solution (see Chapter 1) and v_w, if not zero, must be given as a boundary condition, but what about θ and δ^*? Inspection of Eqs. (2-12) and (2-13) reveals that they involve δ, one of our other desired outputs, but they also involve $u(y)$, which we do not know. Thus, we have one equation and three unknowns! But fluid dynamicists are an imaginative lot; there is a way out of this quandary.

First, however, let us look more closely at θ and δ^*. Equations (2-12) and (2-13) show that they are lengths related to δ, but how? The length δ^* is simpler, so begin with it. The integrand is proportional to $(U_e - u)$, which is the difference between the actual viscous profile and that which would occur if the flow were inviscid; that is, $u = U_e$ for all y down to $y = 0$ (see Fig. 2-2). This is called the *velocity defect* (remember this term, as it is important in turbulent boundary layer analysis). Clearly, there is less mass flow passing through the region $0 \le y \le \delta$ for the viscous profile than there would be if an inviscid profile existed. The difference is proportional to the integral of $(U_e - u)$. We can now ask if it is possible to consider an adjusted inviscid profile that has the same mass flow near the body as the real viscous profile. Such a profile would have the shape shown by the dash/dot curve in Fig. 2-2 such that shaded area I is equal to shaded area II. That would satisfy the criteria, but how could such a profile occur? Only if the solid surface were *displaced* upward a distance δ^* such that

$$\rho U_e \delta^* = \rho \int_0^\delta (U_e - u)\, dy \qquad (2\text{-}13a)$$

The length δ^* is, therefore, called the *displacement thickness*. A parallel argument can be made concerning the flow of momentum through the region $0 \le y \le \delta$ compared to that for an inviscid profile. It turns out that the surface must be shifted upward by an amount θ. The length θ is called the *momentum*

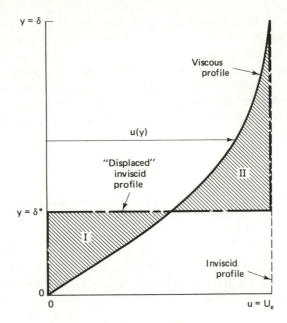

Figure 2-2 Schematic illustration of the definition of the displacement thickness, δ^*.

thickness. These thicknesses are often also used in reporting experimental measurements, since it is difficult accurately to determine δ, the point where $u(x, y)$ blends smoothly into U_e because of scatter in the data. The integrals can be more accurately found, as the scatter tends to cancel out.

2-3 SOLUTION OF THE INTEGRAL MOMENTUM EQUATION

2-3-1 The Pohlhausen Method

At this point it is necessary to face directly the problem posed by one equation [Eq. (2-14)] and three unknowns—C_f, δ^*, and θ. Observe that if $u(y)$ were known, all three quantities would be defined by Eqs. (2-8), (2-12), and (2-13). Of course that is foolish, because if we knew $u(y)$, we would not be bothering with this equation at all. Looking more closely, however, it can be seen that only

$$\frac{u}{U_e} = f\left(\frac{y}{\delta}\right) \qquad (2\text{-}15)$$

that is, only the nondimensional *shape* of the profile, is needed, not $u(y)$ explicitly. If we are bold enough to guess the shape of the profile, the problem will reduce to one equation for one unknown, $\delta(x)$. One is emboldened by the fact that the profile will be integrated for δ^* and θ, so any inaccuracies will tend to cancel. That is correct, but it is also necessary to differentiate the assumed profile to get τ_w, and that operation tends to amplify inaccuracies. In any event, it is instructive to try a simple problem and see how all this works.

The simplest possible boundary layer problem is one where the inviscid solution is $U_e = $ constant, which is flow over a solid flat plate. In that case, Eq. (2-14) becomes

$$\frac{d\theta}{dx} = \frac{C_f}{2} \tag{2-14a}$$

The simplest guess for a profile that meets a minimum of physical criteria is a straight line joining the conditions $u(0) = 0$ and $u(\delta) = U_e$:

$$\frac{u}{U_e} = \frac{y}{\delta} \tag{2-16}$$

Using this in Eq. (2-12) gives $\theta = \delta/6$ and in Eq. (2-8) gives $\tau_w = \mu(\partial u/\partial y)_{y=0} = \mu U_e/\delta$, so Eq. (2-14a) becomes

$$\delta \frac{d\delta}{dx} = \frac{6\,(\mu/\rho)}{U_e} \tag{2-17}$$

which is one equation for the one unknown δ, as promised. This is an easy equation to solve since $U_e = $ constant for this problem. Also, for a flat plate, the boundary layer thickness at the leading edge, $x = 0$, will be zero, so the single required initial condition is just $\delta(0) = 0$. The solution then is

$$\delta(x) = \sqrt{\frac{12\,(\mu/\rho)x}{U_e}} \tag{2-18}$$

which says that δ grows as $\sqrt{x}$. Rearranging, Eq. (2-18) can be written as

$$\frac{\delta}{x} = \sqrt{(12)\frac{\mu}{\rho U_e x}} = 3.46 \mathrm{Re}_x^{-1/2} \tag{2-19}$$

The variation of $C_f(x)$ can be found by substituting Eq. (2-18) into the relation for τ_w above to yield

$$\tau_w = \frac{\mu U_e}{\delta} = \frac{\mu U_e}{\sqrt{12(\mu/\rho)x/U_e}} \tag{2-20}$$

or

$$C_f \equiv \frac{\tau_w}{\frac{1}{2}\rho U_e^2} = 0.577 \mathrm{Re}_x^{-1/2} \tag{2-21}$$

How good is this crude solution? Experiment and elaborate exact analysis of this problem (presented in Chapter 4) indicate that

$$\frac{\delta}{x} = 5.2\text{Re}_x^{-1/2} \tag{2-22}$$

and

$$C_f = 0.664\text{Re}_x^{-1/2} \tag{2-23}$$

Thus, this simple analysis has achieved the correct dependence on Re_x and fairly good numerical values for the coefficients.

Clearly, one could expect to do better with a more realistic profile shape. Pohlhausen (1921b) used the function

$$\frac{u}{U_e} = a + b\left(\frac{y}{\delta}\right) + c\left(\frac{y}{\delta}\right)^2 + d\left(\frac{y}{\delta}\right)^3 + e\left(\frac{y}{\delta}\right)^4 \tag{2-24}$$

The coefficients a to e were determined by forcing satisfaction of a set of physically reasonable conditions on the profile. He used

$$y = 0: \qquad u = 0 \qquad y = \delta: \quad u = U_e$$

$$\mu \frac{\partial^2 u}{\partial y^2} = \frac{dp}{dx} \qquad\qquad \frac{\partial u}{\partial y} = 0 \tag{2-25}$$

$$\frac{\partial^2 u}{\partial y^2} = 0$$

The second condition at $y = 0$ can be developed by considering a small volume of fluid ($dx : dy : 1$) right on the wall. Since $u = v = 0$ on the wall, the momentum flux in and out of such a volume is negligible compared to the forces on the volume. Thus, equilibrium on the volume is achieved simply by a balance of the pressure force and the shear force. The net pressure force is

$$p(dy) - \left(p + \frac{dp}{dx}\,dx\right)dy$$

and the net shear force is

$$-\left(\mu\frac{\partial u}{\partial y}\right)dx + \mu\left(\frac{\partial u}{\partial y} + \frac{\partial}{\partial y}\left(\frac{\partial u}{\partial y}\right)dy\right)dx$$

so the balance produces the stated condition. The result of applying these conditions to Eq. (2-24) is

$$a = 0, \qquad b = 2 + \frac{\lambda}{6}, \qquad c = -\frac{\lambda}{2},$$

$$d = -2 + \frac{\lambda}{2}, \qquad e = 1 - \frac{\lambda}{6} \tag{2-26}$$

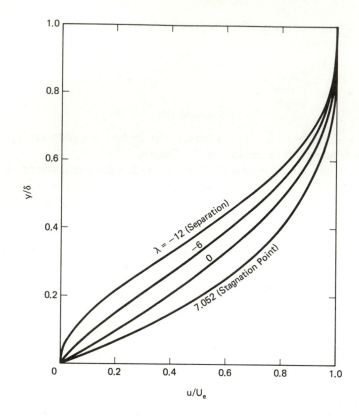

Figure 2-3 Laminar boundary layer velocity profiles by the Pohlhausen method for various values of the pressure gradient.

where $\lambda \equiv (\delta^2/\nu)\, dU_e/dx$ is the *Pohlhausen pressure gradient parameter*. Profiles for various values of λ are shown in Fig. 2-3. The profiles for *favorable* pressure gradients ($dp/dx < 0$ and $\lambda > 0$) (i.e., no tendency toward separation) have a fuller shape than the flat-plate case. For *adverse* pressure gradients ($dp/dx > 0$ and $\lambda < 0$), the fluid near the wall slows down rapidly, leading to an inflection point in the profile. Taking the first derivative of the profile using Eq. (2-26) and evaluating at the wall, we find that $\tau_w = \mu(\partial u/\partial y)_{y=0} = 0$ (i.e., separation) when $\lambda = -12$. Many calculations have been made, and the results, when compared to experiment, indicate that this criterion for separation is rather crude. Attempts to improve the results by using other profiles have not been overly successful. Nonetheless, the *momentum integral method*, as it is called, remains popular for obtaining simple approximate solutions.

2-3-2 The Thwaites–Walz Method

Actually, the calculations with the Pohlhausen method are somewhat bothersome, but fortunately the problem has been reformulated by Thwaites

(1949) and Walz (1941) into a very convenient form. Multiply Eq. (2-14) by $U_e \theta/v$ and rearrange using $H \equiv \delta^*/\theta$ to get

$$\frac{\tau_w \theta}{\mu U_e} = \frac{U_e \theta}{v} \frac{d\theta}{dx} + \frac{\theta^2\, dU_e/dx}{v} (H + 2) \qquad (2\text{-}27)$$

The quantity H is a nondimensional function of the profile shape alone, as is the term on the left-hand side, which is commonly called S, the *shear correlation function*. With Eq. (2-27), it is more convenient to use $\Lambda \equiv (\theta^2/v)\, dU_e/dx$ rather than λ from before. Now, saying that H and S are functions of the profile shape alone is the same as saying $H(\Lambda)$ and $S(\Lambda)$. With this, Eq. (2-27) can be rewritten as

$$U_e \frac{d}{dx}\left[\frac{\Lambda}{dU_e/dx}\right] = 2[S(\Lambda) - \Lambda(H(\Lambda) + 2)] = F(\Lambda) \qquad (2\text{-}28)$$

The Pohlhausen profile or any other or experimental data can be used to find the right-hand side. A good fit to the bulk of the available information is

$$F(\Lambda) = 0.45 - 6.0\Lambda \qquad (2\text{-}29)$$

TABLE 2-1   Shear and Shape
Functions Correlated by
Thwaites (1949)

Λ	$H(\Lambda)$	$S(\Lambda)$
+0.25	2.00	0.500
0.20	2.07	0.463
0.14	2.18	0.404
0.12	2.23	0.382
0.10	2.28	0.359
+0.080	2.34	0.333
0.064	2.39	0.313
0.048	2.44	0.291
0.032	2.49	0.268
0.016	2.55	0.244
0.0	2.61	0.220
−0.016	2.67	0.195
−0.032	2.75	0.168
−0.048	2.87	0.138
−0.064	3.04	0.104
−0.080	3.30	0.056
−0.084	3.39	0.038
−0.088	3.49	0.015
−0.090	3.55 (separation)	0.000

With this, Eq. (2-28) can be integrated once and for all to yield

$$\theta^2(x) = \frac{0.45\nu}{U_e^6} \int_0^x U_e^5(x)\, dx \tag{2-30}$$

This simple expression represents an approximate answer to all low-speed, steady, planar, laminar boundary layer problems!

A given problem is specified by $U_e(x)$ from the inviscid solution. The first step is to solve Eq. (2-30) for $\theta(x)$. Using $U_e(x)$, next find $\Lambda(x) = (\theta^2/\nu)\, dU_e/dx$. We now also require correlations for $S(\Lambda)$ and perhaps $H(\Lambda)$. Thwaites (1949) gave the results listed in Table 2-1. At each point along the surface the appropriate S for the $\Lambda(x)$ is found and then the definition of S is unwound to yield C_f as

$$C_f = \frac{2\mu}{\rho U_e \theta} S(\Lambda) \tag{2-31}$$

Since the solutions from this procedure are based on a wider range of profile shapes than just Eq. (2-24), as well as experiment, they are better than those obtained with the Pohlhausen procedure. The range of inaccuracy is generally less than 10% for favorable pressure gradients, but about 20 to 30% for strong, adverse pressure gradients.

2-3-3 Flows with Suction or Injection

Flows with suction or injection through a porous wall are of practical interest for cooling (with injection of a cool fluid), delaying transition to turbulence by suction (see Chapter 6), and prevention of separation in an adverse pressure gradient by suction. Generally speaking, the momentum integral method is not accurate for these flows, especially those with injection, so detailed discussion of such cases is reserved for later in this text after more powerful methods have been introduced. However, Prandtl (1935) has presented a simple treatment of separation prevention by suction that is illuminating. He considered the case of incipient separation, which with the quartic profile, Eq. (2-24), corresponds to $\lambda = -12$. For that profile, $\delta^*/\delta = \frac{2}{5}$, $\frac{\theta}{\delta} = \frac{4}{35}$, and $\tau_w = \mu(\partial u/\partial y)_{y=0} = 0$. Substituting into the integral momentum equation, Eq. (2-14), and making the assumption that $d\delta/dx = d\delta^*/dx = d\theta/dx = 0$ because the profile is assumed to be maintained constant by the suction, one gets

$$v_w = \frac{22}{35} \delta \frac{dU_e}{dx} \tag{2-32}$$

In Eq. (2-25), we had

$$\left.\frac{\partial^2 u}{\partial y^2}\right|_w = \frac{1}{\mu}\frac{dp}{dx} = -\frac{U_e}{\nu}\frac{dU_e}{dx} \tag{2-33}$$

The profile shape gives

$$\left.\frac{\partial^2 u}{\partial y^2}\right|_w = 12 \frac{U_e}{\delta^2} \tag{2-34}$$

Thus, one can say that

$$\delta = \sqrt{\frac{12v}{-dU_e/dx}} \tag{2-35}$$

and Eq. (2-32) becomes

$$v_w = -2.18 \sqrt{-v \frac{dU_e}{dx}} \tag{2-36}$$

as the suction velocity necessary to maintain incipient separation and prevent flow reversal.

2-4 THE INTEGRAL ENERGY EQUATION

Under the restrictions of planar, steady, constant-density, constant-property flow, there are still cases of important thermal energy, or heat, transfer. Consider 50°C air flowing over a surface at 30°C. There is significant heat transfer, but the range of temperature is small enough to safely neglect variations in the density or the physical properties. A *thermal* boundary layer will develop with a temperature profile varying from $T = T_w(x)$ at $y = 0$ to $T = T_e(x)$ at $y = \delta_T(x)$. The thermal boundary layer thickness δ_T may be thicker or thinner than the velocity boundary layer, depending on the physical properties and the particular problem of interest.

The physical properties combine into the dimensionless *Prandtl number*

$$\mathrm{Pr} \equiv \frac{\mu c_p}{k} \tag{2-37}$$

which represents the ratio of coefficients for diffusion of momentum to diffusion of heat. Interestingly, most common gases have $\mathrm{Pr} \approx 0.70$, which means that heat will diffuse faster than momentum. All other things being equal, then, a thermal boundary layer in a gas flow is likely to be thicker than the velocity boundary layer in a ratio of roughly 1 : 0.7. The Prandtl number for common liquids (water, oil, and so on) is of the order of 10. At the other extreme, the values for liquid metals are of the order of 10^{-3}. Obviously, the Prandtl number of the fluid is of great importance in boundary layer flows with heat transfer.

Approximate solutions at a similar level to those obtained for the velocity boundary layer in this chapter can be obtained by applying conservation of energy to a suitable control volume (see Fig. 2-4). The appropriate statement

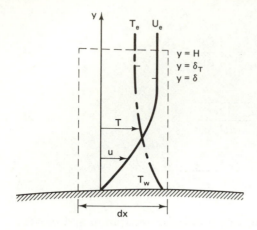

Figure 2-4 Control volume for derivation of the integral energy equation.

of conservation of energy can be written as

thermal energy in + frictional heating

+ heat transfer at the surface

= thermal energy out   (2-38)

The thermal energy carried by the flow into the left-hand side of the control volume is

$$\rho c_p \int_0^H uT \, dy \qquad (2\text{-}39)$$

and that carried out the right-hand side is

$$\rho c_p \int_0^H uT \, dy + \frac{d}{dx}\left[\rho c_p \int_0^H uT \, dy\right] dx \qquad (2\text{-}40)$$

The mass flow through the top surface brings in thermal energy equal to

$$c_p T_e \left\{ \frac{d}{dx}\left[\int_0^H \rho u \, dy\right] dx - \rho_w v_w \, dx \right\} \qquad (2\text{-}41)$$

Since there is viscous shear throughout the boundary layer, $\partial u/\partial y \neq 0$, there is frictional heating. This can be expressed as

$$\int_0^H \mu \left(\frac{\partial u}{\partial y}\right)^2 dy \, dx \qquad (2\text{-}42)$$

There is heat transfer at the surface $-k(\partial T/\partial y)_{y=0}$ and injected fluid carries in energy

$$\rho_w v_w c_p T_w \, dx \qquad (2\text{-}43)$$

Substituting into Eq. (2-38) results in the integral energy equation

$$\frac{d}{dx}\left[\int_0^H (T_e - T)u\, dy\right] + \frac{v}{c_p}\left[\int_0^H \left(\frac{\partial u}{\partial y}\right)^2 dy\right]$$

$$- v_w(T_e - T_w) = \frac{k(\partial T/\partial y)_w}{\rho c_p} \qquad (2\text{-}44)$$

Note that under our assumptions here, $\rho_w = \rho$.

We are now in the same situation as for the integral momentum equation. Unless the shape of the profiles for u and T are specified, there are too many unknowns for this single equation.

2-5 SOLUTION OF THE INTEGRAL ENERGY EQUATION

It should not surprise the reader that most efforts to solve the integral energy equation closely follow the Pohlhausen method; that is, a polynomial assumption for the temperature profile is used in conjunction with a polynomial for the velocity profile.

2-5-1 Unheated Starting-Length Problem

The method can be illustrated and some useful results can be obtained at the same time by considering a specific situation known as the unheated starting-length problem. The flow is as shown in Fig. 2-5 with a solid ($v_w = 0$) flat plate ($U_e = $ constant, $dp/dx = 0$) whose initial portion $0 \le x < x_0$ has the same wall temperature as the static temperature in the free stream (i.e., $q_w = 0$ since there is no temperature difference). For $x \ge x_0$, $T_w \ne T_e$, and there will be heat transfer at the surface. The velocity boundary layer will begin to grow at the leading edge of the plate, $x = 0$, but the thermal boundary layer will only begin at $x = x_0$. Also, it is assumed that the flow speed is low enough to

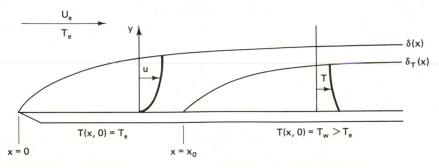

Figure 2-5 Schematic of the flow over a flat plate with an unheated starting length.

neglect viscous heating with respect to the other energy transfer processes. Thus, the second integral in Eq. (2-44) is dropped.

The solution to this problem as given in the heat transfer text by Eckert and Drake (1959) uses cubic polynomials for both the velocity and temperature profiles. For the velocity profile, the condition $\partial^2 u/\partial y^2 = 0$ at $y = \delta$ is dropped from the group in Eq. (2-25). The resulting profile is

$$\frac{u}{U_e} = \frac{3}{2}\left(\frac{y}{\delta}\right) - \frac{1}{2}\left(\frac{y}{\delta}\right)^3 \tag{2-45}$$

The conditions imposed on the temperature profile are

$$y = 0: \quad T = T_w \quad y = \delta_T: \quad T = T_e$$
$$\frac{\partial^2 T}{\partial y^2} = 0 \quad \frac{\partial T}{\partial y} = 0 \tag{2-46}$$

In the same way as for the velocity profile, the second condition on $T(y)$ at $y = 0$ arises from consideration of a small volume of fluid on the wall. Since $u = v = 0$ on the wall, convection can be neglected. Thus, the net energy transfer through that volume,

$$\left(-k\frac{\partial T}{\partial y}\right)dx + k\left(\frac{\partial T}{\partial y} + \frac{\partial}{\partial y}\left(\frac{\partial T}{\partial y}\right)dy\right)dx$$

must be zero, leading to the stated condition. The resulting temperature profile is

$$\frac{T - T_w}{T_e - T_w} = \frac{3}{2}\left(\frac{y}{\delta_T}\right) - \frac{1}{2}\left(\frac{y}{\delta_T}\right)^3 \tag{2-47}$$

Substituting Eqs. (2-45) and (2-47) into Eq. (2-44) and performing the necessary integration and algebraic manipulations leads to

$$U_e(T_e - T_w)\frac{d}{dx}\left[\delta\left(\frac{3}{20}\zeta^2 - \frac{3}{280}\zeta^4\right)\right] = \frac{3}{2}\frac{v(T_e - T_w)}{\text{Pr}(\delta\zeta)} \tag{2-48}$$

where $\zeta \equiv \delta_T/\delta$. We may assume that $\delta_T < \delta$ (see Fig. 2-5) for Pr of order unity and therefore neglect the ζ^4 term with respect to the ζ^2 term.

For a cubic velocity profile, the integral momentum equation for flat-plate flow (Eq. 2-14a) reduces to

$$\delta\frac{d\delta}{dx} = \frac{140}{13}\frac{v}{U_e} \tag{2-49}$$

and the solution is

$$\delta^2 = \frac{280}{13}\frac{vx}{U_e} \tag{2-50}$$

Using these results in Eq. (2-48) yields

$$4x\zeta^2 \frac{d\zeta}{dx} + \zeta^3 = \frac{13}{14}\left(\frac{1}{\mathrm{Pr}}\right) \tag{2-51}$$

The general solution to this first-order, ordinary differential equation is

$$\zeta^3 = cx^{-3/4} + \frac{13}{14}\left(\frac{1}{\mathrm{Pr}}\right) \tag{2-52}$$

The single required initial condition is $\zeta = 0$ (i.e., $\delta_T = 0$) at $x = x_0$, and the final solution may be written

$$\frac{\delta_T}{\delta} = \frac{1}{1.026\mathrm{Pr}^{1/3}}\left[1 - \left(\frac{x_0}{x}\right)^{3/4}\right]^{1/3} \tag{2-53}$$

For $x_0 = 0$ (i.e., when the whole plate is heated),

$$\frac{\delta_T}{\delta} = \frac{1}{1.026\mathrm{Pr}^{1/3}} \tag{2-54}$$

This expression is valid for Pr of order unity despite our earlier assumption of $\delta_T < \delta$.

Of course, the heat transfer rate at the wall is generally of greater practical interest than $\delta_T(x)$. Having $\delta_T(x)$, we can evaluate $q_w = -k(\partial T/\partial y)_{y=0}$ using Eq. (2-47). It is common to express the wall heat transfer in terms of a *film coefficient* h, defined by

$$q_w = -k\left(\frac{\partial T}{\partial y}\right)_{y=0} \equiv h(T_w - T_e) \tag{2-55}$$

This dimensional quantity is usually combined into a dimensionless group known as the *Nusselt number*,

$$\mathrm{Nu} \equiv \frac{hx}{k} \tag{2-56}$$

Thus, for this problem

$$\mathrm{Nu} = 0.332\mathrm{Pr}^{1/3}\left(\frac{\rho U_e x}{\mu}\right)^{1/2}\left[1 - \left(\frac{x_0}{x}\right)^{3/4}\right]^{-1/3} \tag{2-57}$$

In this instance, partly by good fortune, we get the correct coefficient, agreeing with the exact solution for the $x_0 = 0$ case.

Alternatively, the film coefficient can be put in dimensionless form as

$$\mathrm{St} \equiv \frac{h}{\rho U_e c_p} \tag{2-58}$$

called the *Stanton number*. Then the solution becomes

$$St = \frac{Nu}{Re \cdot Pr} = 0.332 Pr^{-2/3} Re^{-1/2} \left[1 - \left(\frac{x_0}{x} \right)^{3/4} \right]^{-1/3} \quad (2\text{-}59)$$

Comparing this result for the case $x_0 = 0$ with Eq. (2-23), we can see that

$$St \cdot Pr^{2/3} = \frac{C_f}{2} \quad (2\text{-}60)$$

showing a direct relation of skin friction to heat transfer called the *Reynolds analogy*.

2-5-2 Nonuniform Wall Temperature

Many practical problems involve cases where the wall temperature varies in a general fashion. The specific example of a step change in wall temperature analyzed above can be generalized for essentially arbitrary variations of $T_w(x)$. The key lies in the fact that Eq. (2-44) (and, for that matter, the corresponding exact boundary layer energy equation considered in Chapter 3) is *linear* in the temperature. That means that the sum of any two (or more) solutions corresponding to two (or more) separate sets of boundary conditions represents an exact solution to the new problem specified by boundary conditions formed by the sum of the two (or more) sets of original boundary conditions. Thus, one says that the solutions may be *superimposed*. Consider the wall temperature variation shown in Fig. 2-6. The actual $T_w(x)$ shown by the solid curve can be approximated by the series of step changes shown dashed. In such a situation,

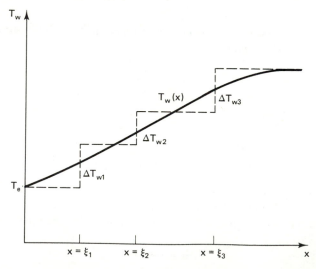

Figure 2-6 Variable wall temperature distribution approximated by a series of step changes.

the total heat flow at a location x is the sum of the contributions of the various step changes

$$q_w = \sum_1^N \hbar(x, \xi_i)(\Delta T_{w_i}) \qquad (2\text{-}61)$$

Here, $\hbar(x, \xi_i)$ represents the film coefficient for heat transfer at x for a single step change, ΔT_{w_i}, located at $x = \xi_i$. This is clearly obtained from Eq. (2-57), writing ξ_i for x_0. This summation can be extended in the limit to an integral for a series of infinitesimal step changes as

$$q_w = \int_0^x \hbar(x, \xi) \frac{dT_w}{d\xi} \, d\xi \qquad (2\text{-}62)$$

A simple approximate procedure for carrying out the required calculations has been developed by Eckert et al. (1957).

2-5-3 Method of Smith and Spalding

Smith and Spalding (1958) extended the ideas behind the Thwaites–Walz method (see Sec. 2-3-2) for the integral momentum equation to the solution of the integral energy equation for cases with $U_e \neq$ constant but $T_w =$ constant. First, the concept of a *conduction thickness* δ_t is introduced as

$$\delta_t \equiv \frac{k(T_w - T_e)}{q_w} \qquad (2\text{-}63)$$

This is combined with the edge velocity gradient to form a dimensionless parameter Λ_t akin to the pressure gradient parameter Λ:

$$\Lambda_t \equiv \frac{\delta_t^2 \, dU_e/dx}{\nu} \qquad (2\text{-}64)$$

By analogy with Eq. (2-28), we seek a correlation of the form

$$U_e \frac{d}{dx}\left[\frac{\Lambda_t}{dU_e/dx}\right] = F_t(\Lambda_t, \text{Pr}) \qquad (2\text{-}65)$$

Smith and Spalding suggest the curve fit

$$F_t = a(\text{Pr}) - b(\text{Pr})\Lambda_t \qquad (2\text{-}66)$$

with

$$a = 9.072(\text{Pr})^{-0.70} \quad \text{and} \quad b = 2.95(\text{Pr})^{0.07} \qquad (2\text{-}67)$$

in the range $0.1 \leq \text{Pr} \leq 10$. For air with $\text{Pr} \approx 0.72$, $a = 11.42$ and $b = 2.88$. Equation (2-65), with Eqs. (2-66) and (2-67), can be integrated to give

$$\delta_t^2 = \frac{a\nu}{U_e^b} \int_0^x U_e^{b-1} \, dx \qquad (2\text{-}68)$$

Looking at Eqs. (2-55) and (2-63), one can see that the film coefficient is related to the conduction thickness by

$$h = \frac{k}{\delta_t} \qquad (2\text{-}69)$$

An expression for the Nusselt number can thus be written as

$$\text{Nu} = \frac{a^{-1/2}(Vx/\nu)^{1/2}\, x^{1/2}}{[(U_e/V)^{-b}\int_0^x (U_e/V)^{b-1}\, dx]^{1/2}} \qquad (2\text{-}70)$$

where V is a suitable characteristic velocity [e.g., U_∞ or $U_e(0)$].

2-6 THE INTEGRAL SPECIES CONSERVATION EQUATION

Retaining the assumptions used throughout this chapter that the flow is planar and steady with constant density and constant physical properties, one can still envision problems with significant mass transfer effects. The derivation of an integral equation for species concentration corresponding to the integral energy and momentum equations follows the same general lines as for those equations.

For simplicity, consider the fluid as a mixture of only two species (e.g., water vapor in air) denoted by subscripts 1 and 2. Referring to Fig. 2-1, we can write an integral conservation equation for fluid 1. For a steady flow, the difference between the mass of that fluid coming into the control volume through the left side and going out the right side must enter through the top and the bottom (if the wall is permeable to fluid 1). The mass of fluid 1 coming in the top can be found from the expression for the total mass coming in the top as

$$c_{1e}\left\{\frac{d}{dx}\left[\int_0^H \rho u\, dy\right] dx - \rho_w v_w\, dx\right\} \qquad (2\text{-}71)$$

where c_{1e} is the mass fraction of fluid 1 in the external flow. The difference between the mass of fluid 1 going out the right side and coming in the left side is

$$\frac{d}{dx}\left[\int_0^H c_1 \rho u\, dy\right] dx \qquad (2\text{-}72)$$

If the mass flux by diffusion of fluid 1 at the wall into the volume is $\dot{m}_{iw}$, conservation requires that

$$\dot{m}_{1w} + c_{1e}\frac{d}{dx}\left[\int_0^H \rho u\, dy\right] - c_{1e}\rho_w v_w + \rho_w v_w c_{1w} = \frac{d}{dx}\left[\int_0^H c_1 \rho u\, dy\right] \qquad (2\text{-}73)$$

or

$$\frac{d}{dx}\left[\int_0^H (c_{1e} - c_1)\rho u \; dy\right] - \rho_w v_w(c_{1e} - c_{1w}) = -\dot{m}_{1w} \qquad (2\text{-}73a)$$

Note that $c_2 \equiv 1.0 - c_1$.

2-7 SOLUTION OF THE INTEGRAL SPECIES CONSERVATION EQUATION

The simplest cases are those with a solid wall, so that $v_w = 0$. The conditions to be imposed on the assumed species concentration profile follow directly from those used for the velocity and temperature profiles.

$$
\begin{array}{llll}
y = 0: & c_1 = c_{1w} & y = \delta_c: & c_1 = c_{1e} \\[2mm]
& \dfrac{\partial^2 c_1}{\partial y^2} = 0 & & \dfrac{\partial c_1}{\partial y} = 0
\end{array}
\qquad (2\text{-}74)
$$

Assuming a cubic profile as was done for the temperature [see Eq. (2-47)], we get

$$\frac{c_1 - c_{1w}}{c_{1e} - c_{1w}} = \frac{3}{2}\left(\frac{y}{\delta_c}\right) - \frac{1}{2}\left(\frac{y}{\delta_c}\right)^3 \qquad (2\text{-}75)$$

Using this with a cubic velocity profile for flow over a flat plate [see Eq. (2-45)], the result can be expressed in terms of a *diffusion film coefficient* $\hbar_D$ defined by

$$\dot{m}_{1w} = -\rho D_{12} \left.\frac{\partial c_1}{\partial y}\right|_{y=0} = \rho \hbar_D(c_{1w} - c_{1e}) \qquad (2\text{-}76)$$

as a *Nusselt number for diffusion*

$$\mathrm{Nu}_{\mathrm{Diff}} = \frac{\hbar_D x}{D_{12}} = 0.332 \mathrm{Re}_x^{1/2} \mathrm{Sc}^{1/3} \qquad (2\text{-}77)$$

where Sc is the Schmidt number ($\equiv v/D_{12}$). Note the close relationship to Eq. (2-57) for $x_0 = 0$ (i.e., a constant-temperature plate case). More will be made of this shortly.

Data for mass transfer from a flat plate in low-speed flow are scarce. In Fig. 2-7, the measurements of Christian and Kezios (1959) for naphthalene sublimation (at a low rate, so $v_w \approx 0$) into air flowing axially over cylinders with $(\delta/R) \ll 1$ (so the effects of axisymmetry compared to a planar case are $\approx 2\%$) are compared to Eq. (2-77). The good agreement is clear.

Another interesting case was treated by Eckert and Lieblein (1949). They studied evaporation or condensation of water vapor from a flat plate in an

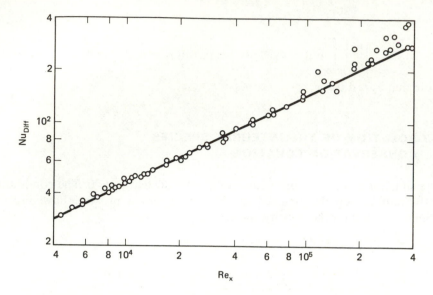

Figure 2-7   Mass transfer rate from a napthalene surface in an airstream compared to analysis. (From Christian and Kezios, 1959.)

airstream. The results depend on two dimensional parameters,

$$A \equiv \frac{D_{12}}{v} \frac{W_1}{W} \quad \text{and} \quad B \equiv \frac{p_{1e} - p_{1w}}{p - p_{1w}} \tag{2-78}$$

which are combined into two others:

$$K_{cp} \equiv 3 - \frac{4 - 2\sqrt{4 - 6B}}{B} \tag{2-79}$$

$$K_c \equiv \frac{3A(2 - \sqrt{4 - 6B})\delta/\delta_c}{A(2 - \sqrt{4 - 6B})\delta/\delta_c - 4}$$

Here p_1 is the partial pressure of fluid 1 and W_1 and W are molecular weights of fluid 1 and the mixture. The ratio δ/δ_c is unity for evaporation of water from the plate, but it increases to nearly two for condensation on the wall with large partial pressure differences.

Velocity and partial pressure profiles are both plotted in Fig. 2-8. Positive values of K_c and K_{cp} indicate evaporation, and negative values indicate condensation. It can be seen that mass transfer has a significant effect on the velocity profile. The value $K_c = K_{cp} = 3$ is for a partial pressure of air equal to zero at the wall, which indicates boiling.

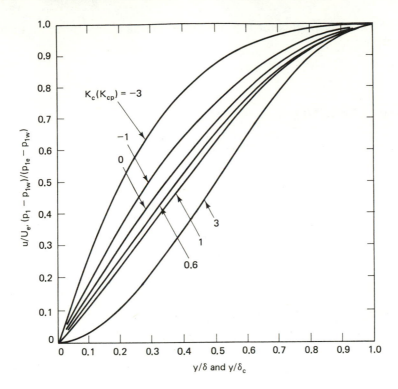

Figure 2-8 Velocity and partial pressure profiles for laminar flow of air and water vapor over a flat surface of evaporating or condensing water. (From Eckert and Lieblein, 1949.)

2-8 RELATIONSHIP OF WALL FRICTION, HEAT TRANSFER, AND MASS TRANSFER

It was noted in Secs. 2-5-1 and 2-7 that the skin friction coefficient and the Stanton number, the Nusselt number, and the Nusselt number for diffusion are related. These observations can be formalized and generalized for cases with $v_w \approx 0$. First, the three processes of shear, heat transfer, and mass transfer are clearly related by virtue of the *laws* that are taken as governing each:

$$\tau = \mu \frac{\partial u}{\partial y}$$

$$q = -k \frac{\partial T}{\partial y} \tag{2-80}$$

$$\dot{m}_i = -\rho D_{ij} \frac{\partial c_i}{\partial y}$$

Also, the film coefficients for heat and mass transfer are defined in a similar manner:

$$q_w = h(T_w - T_e)$$

$$\dot{m}_{iw} = \rho h_D (c_{iw} - c_{ie})$$

(2-81)

Comparing Eq. (2-57) with $x_0 = 0$ and Eq. (2-23), it is clear that

$$\text{Nu} = \text{Re} \cdot \text{Pr}^{1/3} \frac{C_f}{2}$$

(2-82)

Comparing Eqs. (2-77) and (2-23), one can derive

$$\text{Nu}_{\text{Diff}} = \text{Re} \cdot \text{Sc}^{1/3} \frac{C_f}{2}$$

(2-83)

Thus,

$$\frac{h_D}{h} = \frac{D_{12}}{k} \left(\frac{\text{Sc}}{\text{Pr}} \right)^{1/3}$$

(2-84)

which can be rearranged using $\text{Le} = \text{Pr}/\text{Sc}$ to give

$$\frac{h_D}{h} = \frac{\text{Le}^{2/3}}{\rho c_p}$$

(2-85)

2-9 DISCUSSION

In this chapter some aspects of the more widely used approximate techniques for laminar boundary layer analysis have been presented. For the sake of relative simplicity, the coverage has been limited to steady, low-speed, constant-property cases in the planar geometry. Equivalent procedures have also been developed for cases relaxing some or all of these restrictions. Further material is given in subsequent chapters, and the interested reader can refer to older books such as that by Schlichting for more complete historical coverage.

PROBLEMS

2.1. Assume that the velocity profile for flow over a flat plate may be approximated by

$$\frac{u}{U_e} = \tanh \left(2.65 \left(\frac{y}{\delta} \right) \right)$$

and calculate $\delta(x)$ and $C_f(x)$.

2.2. Air at 25°C and 1.0 atm flows over a flat plate at 30 m/s. How thick is the boundary layer at a distance of 3 cm from the leading edge? How thick is the displacement thickness?

2.3. Consider a flat plate in a medium at rest except for a line sink located in the plate surface at $x = L$ as shown in the sketch. Take the inviscid velocity field produced by the sink as $v_r = -C(L/r)$, where r is the radial distance from the sink to any point in the flow. Determine the momentum thickness distribution $\theta(x)$ for $0 \le x \le L$.

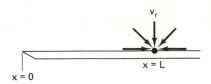

2.4. Using a linear profile for the velocity, Eq. (2-16), and a cubic profile for the temperature, Eq. (2-47), develop an expression for Nu(x) equivalent to Eq. (2-57).

2.5. For a flat plate heated over its entire length, how is the average film coefficient up to a station x related to the local value at the same station?

2.6. Water at 30°C flows over a flat plate at 0.5 m/s. The plate is heated to 40°C starting at 1 cm from the leading edge. What is the total heat transfer from the plate up to a distance 5 cm from the leading edge?

2.7. Hydrogen at 500°C and 15 atm flows over a plate at 2 m/s. The first 2 cm of the plate is maintained at 500°C, the next 3 cm is at 600°C, and the rest of the plate is at 650°C. What is the value of the heat transfer rate at 6 cm from the leading edge?

2.8. Calculate the heat transfer rate at $x = L/2$ for airflow in the pattern of Prob. 2.3 with $C = 5$ m/s for $T_w = 100$°C and $T_e = 75$°C at 1 atm.

2.9. Repeat Prob. 2.8 but with water as the fluid and $C = 0.5$ m/s.

2.10. What is the value of the film coefficient for diffusion h_D at a distance of 0.5 m for a mixture of CO_2 and air flowing over a flat plate at 3 m/s at 1 atm? Assume a small concentration of CO_2 such that all the properties such as density, viscosity, and specific heat may be taken as those for pure air. How does h_D compare to the heat transfer film coefficient h?

2.11. An airstream at 20°C and 1.0 atm is flowing at 3 m/s over a flat water surface at 15°C. The humidity of the air is such that the partial pressure of water vapor is 0.005 atm. Assuming that the water vapor at the liquid/gas interface is saturated (partial pressure of 0.017 atm at 15°C), use Eqs. (2-76) and (2-77) to calculate the evaporation rate of the water at a distance of 0.1 m from the leading edge.

3

DIFFERENTIAL EQUATIONS
OF MOTION
FOR LAMINAR FLOW

3-1 INTRODUCTION

In this chapter attention is directed to the exact, laminar boundary layer form
of the equations of motion. These are derived by applying the conservation
principles to a differential volume of fluid, and coverage is restricted to planar
or axisymmetric flows of ideal fluids (gases obeying the ideal gas law and
incompressible liquids). The resulting system of equations, although mathematically complex, is capable of describing the fine details of the flow in the
boundary layer. In light of the discussions in Chapter 2, the reader may well
ask why it is worthwhile to go to this level of complexity. First, there are some
practical situations where fine details and/or great precision on gross quantities such as C_f are required. This can be illustrated using a reentry vehicle as
an example. Such devices are designed at the limits of performance and sur-
vivability, so the skin friction drag and wall heat transfer rate must be known
quite accurately. A precision of 1 to 2% is sought. However, it is also important to know the maximum static temperature in the boundary layer, since
that is crucial in predicting the communications *blackout* period. Second, it is
important to have essentially exact, so-called *benchmark*, solutions to some
problems, so that the adequacy of various approximate methods can be
judged. Also, the widespread availability of large digital computers has put the
solution of the exact equations within the reach of most practicing engineers.
It is a matter of history in high technology that as soon as more accurate
solutions become achievable, all workers want to have them. There is little
nostalgia in technology for the old ways of doing things. The fourth and most
important reason for studying the differential equations for the boundary layer

is that approximate methods for turbulent flows have been less successful than for laminar flows. Thus, modern treatments of turbulent flows are based on differential formulations, and the study of the corresponding laminar cases provides a good background for undertaking the turbulent analyses.

3-2 CONSERVATION OF MASS: THE CONTINUITY EQUATION

The derivation of the continuity equation is relatively simple, and the student has probably seen it before. Note that this equation is not affected by an assumption of inviscid flow compared to its form for a real, viscous fluid.

We begin by referring to Fig. 3-1, which shows the planar geometry. The mass entering the left-hand side of the differential volume $(dx : dy : 1)$, assuming a unit depth in the z direction, is

$$\rho u \, dy \qquad (3\text{-}1)$$

and the mass leaving the right-hand side can be written

$$\left(\rho + \frac{\partial \rho}{\partial x} \, dx \right) \left(u + \frac{\partial u}{\partial x} \, dx \right) dy \qquad (3\text{-}2)$$

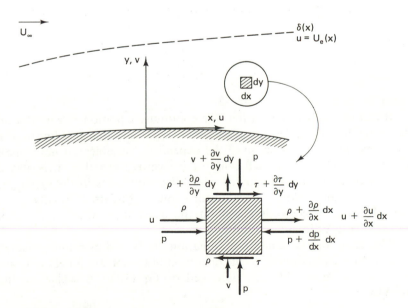

Figure 3-1 Differential volume for conservation of mass and momentum in the boundary layer.

Similarly, the mass flowing in through the bottom is

$$\rho v \, dx \tag{3-3}$$

and that out the top is

$$\left(\rho + \frac{\partial \rho}{\partial y} \, dy\right)\left(v + \frac{\partial v}{\partial y} \, dy\right) dx \tag{3-4}$$

Any net difference between the inflows and outflows must appear as an unsteady change in the mass of fluid in the elemental volume

$$\left(\frac{\partial \rho}{\partial t}\right) dx \, dy \tag{3-5}$$

Combining Eqs. (3-1) to (3-5) yields

$$\frac{\partial \rho}{\partial t} + \frac{\partial(\rho u)}{\partial x} + \frac{\partial(\rho v)}{\partial y} = 0 \tag{3-6}$$

For an axisymmetric flow, the distance from the axis r enters, and the corresponding equation becomes

$$\frac{\partial \rho}{\partial t} + \frac{1}{r}\frac{\partial(\rho u r)}{\partial x} + \frac{1}{r}\frac{\partial(\rho v r)}{\partial r} = 0 \tag{3-7}$$

For constant-density flow, these equations simplify to

$$\frac{\partial u}{\partial x} + \frac{\partial v}{\partial y} = 0 \tag{3-6a}$$

and

$$\frac{1}{r}\frac{\partial(ur)}{\partial x} + \frac{1}{r}\frac{\partial(vr)}{\partial r} = 0 \tag{3-7a}$$

whether the flow is *steady or unsteady.*

It can be easily observed that the continuity equation is a partial differential equation [two independent variables (x, y)] and that it is linear, at least in the constant-density case, since the dependent variables do not appear as products of themselves or each other or derivatives of either. It is also clear that this equation must be coupled with other equations in the system, since there are too many unknowns for this one equation by itself. Finally, as noted before, the fluid viscosity is absent in the exact form whether the fluid is assumed viscous or inviscid.

The form of the continuity equation in two dimensions—planar or axisymmetric—leads directly to the definition of the stream function. Consider the planar, incompressible case described by Eq. (3-6a). A scalar function ψ defined by

$$\frac{\partial \psi}{\partial y} = u; \qquad -\frac{\partial \psi}{\partial x} = v \tag{3-8}$$

will always satisfy Eq. (3-6a), as can be seen by simple substitution. For the axisymmetric geometry, we can have another function Ψ:

$$\frac{\partial \Psi}{\partial r} = ur; \qquad -\frac{\partial \Psi}{\partial x} = vr \qquad (3\text{-}9)$$

If the flow is compressible, one must add the restriction of steady flow, but then we can have

$$\frac{\partial \psi}{\partial y} = \rho u; \qquad -\frac{\partial \psi}{\partial x} = \rho v \qquad (3\text{-}10)$$

or

$$\frac{\partial \Psi}{\partial r} = \rho ur; \qquad -\frac{\partial \Psi}{\partial x} = \rho vr \qquad (3\text{-}11)$$

The concept of a stream function often proves useful in analysis, since it reduces the number of unknowns by one, $(u, v) \rightarrow \psi$ or Ψ, and the continuity equation is automatically solved, so the system of equations is correspondingly reduced by one.

3-3 CONSERVATION OF MOMENTUM: THE MOMENTUM EQUATION

Since the boundary layer approximation is equivalent to the statement that $\partial p/\partial y \approx 0$ across the layer, it is necessary to treat conservation of momentum only in the streamwise, x, direction in detail. Consider again the elemental volume in Fig. 3-1 and note that the detailed derivation given here is only for the planar geometry. The mass entering the left-hand side, $\rho u \, dy$, brings with it x momentum in the amount

$$u(\rho u) \, dy \qquad (3\text{-}12)$$

and the x momentum leaving the right-hand side is

$$\left(u + \frac{\partial u}{\partial x} \, dx\right)\left(\rho u + \frac{\partial(\rho u)}{\partial x} \, dx\right) dy \qquad (3\text{-}13)$$

The mass entering the bottom face $\rho v \, dx$ brings with it x momentum of

$$u(\rho v) \, dx \qquad (3\text{-}14)$$

and the x momentum leaving the top is

$$\left(u + \frac{\partial u}{\partial y} \, dy\right)\left(\rho v + \frac{\partial(\rho v)}{\partial y} \, dy\right) dx \qquad (3\text{-}15)$$

The x momentum in the volume

$$(\rho u) \, dx \, dy \qquad (3\text{-}16)$$

can also change in an unsteady fashion, and that can be expressed simply as

$$\frac{\partial(\rho u)}{\partial t}\, dx\, dy \tag{3-17}$$

The net change in x momentum in the volume is balanced by the resultant of the forces in the x direction.

There are, in general, two kinds of fluid forces: *body forces* and *surface forces*. Body forces act on the bulk of the fluid in a volume as a whole. The most common example is gravity, but under special circumstances magnetic or electric fields can also produce body forces. In most fluid flow problems, except those at very low speed, the influence of body forces is negligible compared to inertia and the other forces in the flow. We will carry along a generalized body force per unit volume f_i in the derivations in this section, but no problems involving body forces will be treated. The interested reader is referred to the chapter "Free Convection" in the heat transfer text by Eckert and Drake (1959), for example, for specific material on that subject.

The primary fluid forces in many practical problems are two surface forces, pressure and viscous forces. These forces influence the fluid in a volume by acting on the surface of the volume. The summation of pressure forces in the x direction is

$$p\, dy - \left(p + \frac{\partial p}{\partial x}\, dx\right) dy \tag{3-18}$$

The only components of the viscous shear that produce forces in the x direction are those acting on the top and bottom of the volume. Considering the typical shape of a velocity profile (see Fig. 2-1), the shear acting on the bottom surface tends to retard the flow, and it can be written

$$-\tau\, dx \tag{3-19}$$

The viscous shear on the top surface tends to accelerate the flow:

$$\left(\tau + \frac{\partial \tau}{\partial y}\, dy\right) dx \tag{3-20}$$

Setting the net change in x momentum equal to the summation of fluid forces in the x direction results in an equation that can be rearranged and simplified after dividing through by $(dx)(dy)$ to give

$$u\left[\frac{\partial \rho}{\partial t} + \frac{\partial(\rho u)}{\partial x} + \frac{\partial(\rho v)}{\partial y)}\right] + \left[\left(\frac{\partial u}{\partial x}\right)\left(\frac{\partial(\rho u)}{\partial x}\right) dx + \left(\frac{\partial u}{\partial y}\right)\left(\frac{\partial(\rho v)}{\partial y}\right) dy\right]$$

$$+ \rho\left[\frac{\partial u}{\partial t} + u\frac{\partial u}{\partial x} + v\frac{\partial u}{\partial y}\right] = -\frac{\partial p}{\partial x} + \frac{\partial \tau}{\partial y} + f_x \tag{3-21}$$

The group of terms in the first set of brackets should look familiar. It is the continuity equation [see Eq. (3-6)]; thus, it is identically equal to zero.

Looking at the terms in the second set of brackets, we see that they, and they alone, still contain the differential lengths, dx and dy. Since this is a differential equation, it remains valid in the limit as dx and $dy \to 0$, so those terms disappear.

The boundary layer momentum equation then becomes

$$\frac{\partial u}{\partial t} + u\frac{\partial u}{\partial x} + v\frac{\partial u}{\partial y} = -\frac{1}{\rho}\frac{\partial p}{\partial x} + \frac{1}{\rho}\frac{\partial \tau}{\partial y} + \frac{f_x}{\rho} \qquad (3\text{-}22)$$

This is valid for planar, unsteady, compressible boundary layer flow with body forces. Note that we have not yet (in all of Chapter 3) had to say that the flow is laminar. Actually, Eq. (3-22) is equally valid for turbulent flow. The only distinction between the two cases will be in how the shear, τ, is *modeled*.

3-3-1 Modeling of the Laminar Shear Stress

This matter can be made clearer by first considering a restricted case—steady, constant-density flow, without body forces. The conclusions reached, however, are general and not influenced by those restrictions. A problem is specified by the appropriate forms of the continuity equation

$$\frac{\partial u}{\partial x} + \frac{\partial v}{\partial y} = 0 \qquad (3\text{-}6a)$$

and the momentum equation

$$u\frac{\partial u}{\partial x} + v\frac{\partial u}{\partial y} = -\frac{1}{\rho}\frac{dp}{dx} + \frac{1}{\rho}\frac{\partial \tau}{\partial y} \qquad (3\text{-}23)$$

and an equation of state, which for this situation is simply $\rho = $ constant. Keeping in mind that the pressure, $p(x)$, is *imposed* on the boundary layer by the inviscid solution, one can see that $p(x)$ is not really a dependent variable in this system. The actual dependent variables are u, v, τ, and ρ. But there are only three equations. One cannot hope to make the situation for ρ much simpler, and it is unreasonable to try to find some direct relation between u and v, so attention is naturally directed to τ.

To close the system mathematically, the variable τ must be related to the other variables in a way that has some general validity, or the whole mathematical structure will be severely restricted. The process is called *modeling*. Fortunately, for laminar flow, the simple expression

$$\tau = \mu\frac{\partial u}{\partial y} \qquad (3\text{-}24)$$

is valid for a wide range of fluids. In that case, $\mu = \mu(T)$ is a physical property of the fluid composition and temperature alone and not dependent on the fluid motion. The matter is vastly more complicated for turbulent flow, as we shall see in Chapter 7.

3-3-2 Forms of the Momentum Equation
for Laminar Flow

Using Eq. (3-24) in Eq. (3-22) yields

$$\rho\left[\frac{\partial u}{\partial t} + u\frac{\partial u}{\partial x} + v\frac{\partial u}{\partial y}\right] = -\frac{\partial p}{\partial x} + \frac{\partial}{\partial y}\left(\mu\frac{\partial u}{\partial y}\right) \tag{3-25}$$

Note that μ must be kept inside the derivative on the right-hand side if the temperature is varying, because $\mu = \mu(T)$.

If the flow is axisymmetric, the distance from the axis enters, and the equation becomes

$$\rho\left[\frac{\partial u}{\partial t} + u\frac{\partial u}{\partial x} + v\frac{\partial u}{\partial r}\right] = -\frac{\partial p}{\partial x} + \frac{1}{r}\frac{\partial}{\partial r}\left(\mu r\left(\frac{\partial u}{\partial r}\right)\right) \tag{3-26}$$

This is to be used with Eq. (3-7).

In most practical instances in the axisymmetric geometry, one is interested in the boundary layer on an axisymmetric body as shown in Fig. 3-2. If a so-called *body system of coordinates* is employed, the appropriate forms of the momentum and continuity equations become

$$\rho\left[\frac{\partial u}{\partial t} + u\frac{\partial u}{\partial x} + v\frac{\partial u}{\partial y}\right] = -\frac{\partial p}{\partial x} + \frac{\partial}{\partial y}\left(\mu\frac{\partial u}{\partial y}\right) \tag{3-27}$$

and

$$r_0^j\frac{\partial \rho}{\partial t} + \frac{\partial(\rho u r_0^j)}{\partial x} + \frac{\partial(\rho v r_0^j)}{\partial y} = 0 \tag{3-28}$$

Where $r_0(x)$ is the local body radius. These equations are valid for $\delta \ll r_0$ and no sharp corners; that is, $d^2 r_0/dx^2$ must be well behaved. Note that Eq. (3-27) now has precisely the same form as the momentum equation for planar flow, Eq. (3-25). Thus, a convenient, unified presentation can be written. Equations

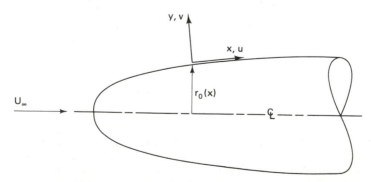

Figure 3-2 Schematic of the body coordinate system for the boundary layer on a body of revolution.

(3-27) and (3-28) hold for flow over bodies in both the planar geometry ($j \equiv 0$) and the axisymmetric geometry ($j \equiv 1$).

If the flow has constant density, the momentum equation shows no explicit change in form—the coefficients involving ρ simply become constant. The continuity equation becomes

$$\frac{\partial(ur_0^j)}{\partial x} + \frac{\partial(vr_0^j)}{\partial y} = 0 \qquad (3\text{-}28a)$$

for both steady and unsteady flow. If the assumption of constant properties is also made, which is common with the constant-density assumption for gas flows, then the momentum equation becomes

$$\frac{\partial u}{\partial t} + u\frac{\partial u}{\partial x} + v\frac{\partial u}{\partial y} = -\frac{1}{\rho}\frac{\partial p}{\partial x} + v\frac{\partial^2 u}{\partial y^2} \qquad (3\text{-}27a)$$

For steady flow, the first term on the left-hand side is dropped.

Another useful form of the momentum equation for planar or axisymmetric flows is developed by the introduction of a stream function as the dependent variable in place of the two velocity components (u, v). This can be illustrated for the planar, constant-density, and property case. A stream function defined by Eq. (3-8) is substituted into Eq. (3-27a) to yield

$$\frac{\partial^2 \psi}{\partial y \, \partial t} + \frac{\partial \psi}{\partial y}\frac{\partial^2 \psi}{\partial y \, \partial x} - \frac{\partial \psi}{\partial x}\frac{\partial^2 \psi}{\partial y^2} = -\frac{1}{\rho}\frac{\partial p}{\partial x} + v\frac{\partial^3 \psi}{\partial y^3} \qquad (3\text{-}29)$$

This is one equation for the one unknown, $\psi(x, y, t)$, with the equation of state, $\rho = $ constant, implied.

3-4 CONSERVATION OF ENERGY: THE ENERGY EQUATION

The thermal energy in the flow can be expressed in terms of the internal energy, the enthalpy, or the temperature. Also, either static or stagnation (or *total*) versions of any of these three can be employed. For this reason, more different forms of the energy equation appear in the literature than for any of the other equations. In this book, we will work primarily with the form based on the static temperature. In any event, each of these dependent variables is a scalar, so only a single equation is required.

Conservation of energy is expressed by the first law of thermodynamics, which states that the change in energy of a system results from the difference between heat transfer and work. Consider each process separately, referring to Fig. 3-3. We retain the boundary layer assumptions on the flow: $\partial p/\partial y \approx 0$ and $\partial u/\partial y \gg \partial u/\partial x$ and $u \gg v$ and use the equivalent assumption for the temperature field $\partial T/\partial y \gg \partial T/\partial x$. Changes in the energy of the flow are expressed in terms of changes in the complete internal energy—thermal $e(T)$ plus

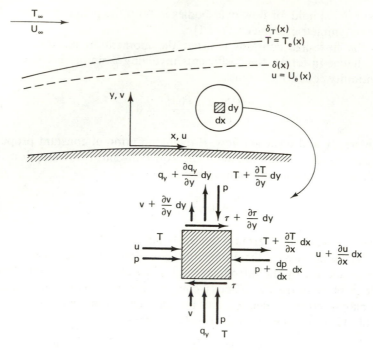

Figure 3-3 Differential volume for conservation of energy in the boundary layer.

kinetic—as

$$de + d\left(\frac{u^2}{2}\right) = c_v\, dT + d\left(\frac{u^2}{2}\right) \tag{3-30}$$

for a perfect gas. In this section we consider a fluid of only one chemical species.

The mass flow in the left-hand side brings in energy at a rate

$$\rho u\left(e + \frac{u^2}{2}\right) dy \tag{3-31}$$

Mass flow out the right-hand side carries out energy at a rate

$$\left[\rho u + \frac{\partial(\rho u)}{\partial x}\, dx\right]\left[\left(e + \frac{u^2}{2}\right) + \frac{\partial}{\partial x}\left(e + \frac{u^2}{2}\right) dx\right] dy \tag{3-32}$$

For the bottom and top surfaces, one obtains

$$\rho v\left(e + \frac{u^2}{2}\right) dx \tag{3-33}$$

and

$$\left[\rho v + \frac{\partial(\rho v)}{\partial y}\,dy\right]\left[\left(e + \frac{u^2}{2}\right) + \frac{\partial}{\partial y}\left(e + \frac{u^2}{2}\right)dy\right]dx \qquad (3\text{-}34)$$

There can also be an unsteady change in the energy in the volume expressed as

$$\frac{\partial}{\partial t}\left[\rho\left(e + \frac{u^2}{2}\right)\right]dx\,dy \qquad (3\text{-}35)$$

Equations (3-31) to (3-35) can be combined using the continuity equation, as was done for the momentum equation with Eqs. (3-12) to (3-17), to form the total rate of change of the energy in the volume as

$$\rho\left[\frac{\partial}{\partial t}\left(e + \frac{u^2}{2}\right) + u\frac{\partial}{\partial x}\left(e + \frac{u^2}{2}\right) + v\frac{\partial}{\partial y}\left(e + \frac{u^2}{2}\right)\right] \qquad (3\text{-}36)$$

Denote the heat flux vector as q_i. Within the boundary layer approximation, the only important terms are the heat flow in the bottom $q_y(dx)$ and out the top,

$$\left(q_y + \frac{\partial q_y}{\partial y}\,dy\right)dx \qquad (3\text{-}37)$$

So the net heat flow rate to the volume is

$$-\frac{\partial q_y}{\partial y}\,dy\,dx \qquad (3\text{-}38)$$

Work can be done on the system by all the forces that are acting. Neglecting body forces, one need only be concerned with pressure and viscous forces. Work is a force acting through a distance, so for a flowing system, the work per unit time becomes a force times a velocity. The pressure forces do net work on the side surfaces in the amount

$$-\frac{\partial(pu)}{\partial x}\,dx\,dy \qquad (3\text{-}39)$$

and on the top and bottom surfaces

$$-\frac{\partial(pv)}{\partial y}\,dy\,dx \qquad (3\text{-}40)$$

In both of these equations, one must carefully reckon the signs, considering the orientation of the pressure force and the velocity components to the surfaces in question.

With the boundary layer approximation, one neglects shear on the sides compared to those on the top and bottom surfaces, so we have for the work done by frictional forces

$$-\tau u\ dx + \left(\tau u + \frac{\partial(\tau u)}{\partial y}\ dy\right) dx \qquad (3\text{-}41)$$

Combining Eqs. (3-39) to (3-41), the net work per unit time is obtained as

$$\frac{\partial(\tau u)}{\partial y} - \frac{\partial(pu)}{\partial x} - \frac{\partial(pv)}{\partial y} \qquad (3\text{-}42)$$

Now combine Eq. (3-36), which is the rate of increase of total energy, with Eq. (3-38), which is the rate at which heat is transferred into the element, and Eq. (3-42), which is the rate at which the stresses on the boundaries do work on the volume through the first law to get

$$\rho\left[\frac{\partial}{\partial t}\left(e + \frac{u^2}{2}\right) + u\frac{\partial}{\partial x}\left(e + \frac{u^2}{2}\right) + v\frac{\partial}{\partial y}\left(e + \frac{u^2}{2}\right)\right]$$

$$= -\frac{\partial q_y}{\partial y} + \frac{\partial(\tau u)}{\partial y} - \frac{\partial(pu)}{\partial x} - \frac{\partial(pv)}{\partial y} \qquad (3\text{-}43)$$

This is a correct, but inconvenient, form of the energy equation, so we will make some substitutions.

First, replace the internal energy by the enthalpy defined as

$$h \equiv e + \frac{p}{\rho}; \qquad dh = c_p\ dT \qquad (3\text{-}44)$$

The operator on the left-hand side of Eq. (3-43) can be written for shorthand as

$$\rho\left[\frac{\partial(\cdot)}{\partial t} + u\frac{\partial(\cdot)}{\partial x} + v\frac{\partial(\cdot)}{\partial y}\right] = \rho\frac{D(\cdot)}{Dt} \qquad (3\text{-}45)$$

With Eq. (3-44), one can write

$$\rho\frac{De}{Dt} = \rho\frac{D}{Dt}\left(h - \frac{p}{\rho}\right) = \rho\frac{Dh}{Dt} - \frac{Dp}{Dt} + \frac{p}{\rho}\frac{D\rho}{Dt} \qquad (3\text{-}46)$$

Now the continuity equation, Eq. (3-6), can be expanded and rewritten as

$$\frac{D\rho}{Dt} = -\rho\left(\frac{\partial u}{\partial x} + \frac{\partial v}{\partial y}\right) \qquad (3\text{-}6b)$$

Using this and rearranging Eq. (3-46), there results

$$\rho\frac{De}{Dt} = \rho\frac{Dh}{Dt} - \frac{\partial(pu)}{\partial x} - \frac{\partial(pv)}{\partial y} - \frac{\partial p}{\partial t} \qquad (3\text{-}47)$$

Thus, Eq. (3-43) can be recast to read

$$\rho\left[\frac{\partial}{\partial t}\left(h + \frac{u^2}{2}\right) + u\frac{\partial}{\partial x}\left(h + \frac{u^2}{2}\right) + v\frac{\partial}{\partial y}\left(h + \frac{u^2}{2}\right)\right]$$

$$= -\frac{\partial q_y}{\partial y} + \frac{\partial(\tau u)}{\partial y} + \frac{\partial p}{\partial t} \qquad (3\text{-}48)$$

The second step necessary to achieve the desired form of the energy equation is to remove the $u^2/2$ terms. For this purpose, consider the momentum equation, Eq. (3-22), dropping the body force term and multiplying through by the streamwise velocity component u, giving

$$\rho\left[\frac{\partial(u^2/2)}{\partial t} + u\frac{\partial(u^2/2)}{\partial x} + v\frac{\partial(u^2/2)}{\partial y}\right] = -u\frac{\partial p}{\partial x} + u\frac{\partial \tau}{\partial y} \qquad (3\text{-}49)$$

This is often called the *mechanical energy* equation. Now, subtract Eq. (3-49) from (3-48) and obtain

$$\rho\left[\frac{\partial h}{\partial t} + u\frac{\partial h}{\partial x} + v\frac{\partial h}{\partial y}\right] = u\frac{\partial p}{\partial x} - \frac{\partial q_y}{\partial y} + \frac{\partial p}{\partial t} + \tau\frac{\partial u}{\partial y} \qquad (3\text{-}50)$$

or, using Eq. (3-44),

$$\rho c_p\left[\frac{\partial T}{\partial t} + u\frac{\partial T}{\partial x} + v\frac{\partial T}{\partial y}\right] = u\frac{\partial p}{\partial x} + \frac{\partial p}{\partial t} - \frac{\partial q_y}{\partial y} + \tau\frac{\partial u}{\partial y} \qquad (3\text{-}50a)$$

This is the desired result—the exact, boundary layer energy equation written in terms of the static temperature. It is valid for both laminar and turbulent flow, since the shear τ and now also the heat transfer rate q_y have not been *modeled*.

3-4-1 Modeling of the Laminar Heat Flux

For laminar flow of many common fluids, a suitable approximation is

$$q_y = -k\frac{\partial T}{\partial y}$$

where $k = k(T)$ is a physical property of the fluid composition and temperature. It does not depend on the fluid motion. Using this expression and that for the shear in laminar flow, the energy equation becomes

$$\rho c_p\left[\frac{\partial T}{\partial t} + u\frac{\partial T}{\partial x} + v\frac{\partial T}{\partial y}\right] = \frac{\partial}{\partial y}\left(k\frac{\partial T}{\partial y}\right) + \mu\left(\frac{\partial u}{\partial y}\right)^2 + u\frac{\partial p}{\partial x} + \frac{\partial p}{\partial t} \qquad (3\text{-}51)$$

For constant-property flow, the energy equation can be written

$$\rho c_p\left[\frac{\partial T}{\partial t} + u\frac{\partial T}{\partial x} + v\frac{\partial T}{\partial y}\right] = k\frac{\partial^2 T}{\partial y^2} + \mu\left(\frac{\partial u}{\partial y}\right)^2 + u\frac{\partial p}{\partial x} + \frac{\partial p}{\partial t} \qquad (3\text{-}52)$$

If the fluid is also incompressible, we can write

$$\rho c_v \left[\frac{\partial T}{\partial t} + u \frac{\partial T}{\partial x} + v \frac{\partial T}{\partial y} \right] = k \frac{\partial^2 T}{\partial y^2} + \mu \left(\frac{\partial u}{\partial y} \right)^2 \qquad (3\text{-}53)$$

The apparent simplification achieved in Eq. (3-53) over (3-52) results from starting with Eq. (3-43), subtracting the mechanical energy equation, Eq. (3-49), and then using the continuity equation for incompressible flow, Eq. (3-6a).

Finally, it is important to note that Eq. (3-51) [or (3-52) or (3-53)] holds for flow over axisymmetric bodies if (x, y) are chosen as in Fig. 3-2 in the same way as for the momentum equation.

For axisymmetric flows, in general, the exact forms are

$$\rho c_p \left[\frac{\partial T}{\partial t} + u \frac{\partial T}{\partial x} + v \frac{\partial T}{\partial r} \right] = \frac{1}{r} \frac{\partial}{\partial r} \left(kr \frac{\partial T}{\partial r} \right) + \mu \left(\frac{\partial u}{\partial r} \right)^2 + u \frac{\partial p}{\partial x} + \frac{\partial p}{\partial t} \qquad (3\text{-}54)$$

and

$$\rho c_v \left[\frac{\partial T}{\partial t} + u \frac{\partial T}{\partial x} + v \frac{\partial T}{\partial r} \right] = \frac{k}{r} \frac{\partial}{\partial r} \left(r \frac{\partial T}{\partial r} \right) + \mu \left(\frac{\partial u}{\partial r} \right)^2 \qquad (3\text{-}55)$$

3-5 CONSERVATION OF MASS OF SPECIES: THE SPECIES CONTINUITY EQUATION

The application of the principle of conservation of mass of a species i requires only a slight generalization of the development for the conservation of the total mass in Sec. 3-2. Consider again the differential control volume in Fig. 3-1. The net mass flow of species i between that in the left side and out the right side is

$$\frac{\partial}{\partial x} (\rho u c_i) \, dx \, dy \qquad (3\text{-}56)$$

For the flows in the bottom and out the top, we must now allow for a mass flux of species i by diffusion as well as by convection. Denote the diffusive flux as $\dot{m}_i$. The net mass flow of species i in the y direction is then

$$\frac{\partial}{\partial y} (\rho v c_i + \dot{m}_i) \, dy \, dx \qquad (3\text{-}57)$$

Any accumulation of mass of species i in the volume must appear as an unsteady term

$$\frac{\partial}{\partial t} (\rho c_i) \, dx \, dy \qquad (3\text{-}58)$$

Applying conservation of mass of species i, one obtains after using the continuity equation (for the total mass), Eq. (3-6),

$$\rho\left[\frac{\partial c_i}{\partial t} + u\frac{\partial c_i}{\partial x} + v\frac{\partial c_i}{\partial y}\right] = -\frac{\partial \dot{m}_i}{\partial y} \qquad (3\text{-}59)$$

This holds for laminar or turbulent flow.

3-5-1 Modeling of Laminar Diffusion

Assuming a binary mixture and neglecting thermal and pressure diffusion, Fick's law gives

$$\dot{m}_i = -\rho D_{ij}\frac{\partial c_i}{\partial y} \qquad (3\text{-}60)$$

Thus, Eq. (3-59) becomes

$$\rho\left[\frac{\partial c_i}{\partial t} + u\frac{\partial c_i}{\partial x} + v\frac{\partial c_i}{\partial y}\right] = \frac{\partial}{\partial y}\left(\rho D_{ij}\frac{\partial c_i}{\partial y}\right) \qquad (3\text{-}61)$$

If the fluid has constant density and properties, the equation simplifies to

$$\frac{\partial c_i}{\partial t} + u\frac{\partial c_i}{\partial x} + v\frac{\partial c_i}{\partial y} = D_{ij}\frac{\partial^2 c_i}{\partial y^2} \qquad (3\text{-}62)$$

For axisymmetric flows, we have

$$\rho\left[\frac{\partial c_i}{\partial t} + u\frac{\partial c_i}{\partial x} + v\frac{\partial c_i}{\partial r}\right] = \frac{1}{r}\frac{\partial}{\partial r}\left(\rho D_{ij} r\frac{\partial c_i}{\partial r}\right) \qquad (3\text{-}63)$$

3-5-2 Energy Transfer by Mass Transfer

Mass transfer can contribute to energy transfer even if the fluid is at constant temperature. This is a consequence of the fact that different species generally have different specific heats (e.g., c_p). For a laminar flow, this results in an additional term in the energy equation expressed as a function of stagnation enthalpy, Eq. (3-48), as

$$-\frac{\partial}{\partial y}\left[\left(\frac{1}{\text{Le}} - 1\right)\rho D_{ij}\sum_i h_i\frac{\partial c_i}{\partial y}\right] \qquad (3\text{-}64)$$

where

$$h_i = \int c_{pi}\, dT \qquad (3\text{-}65)$$

The energy equation for problems with mass transfer is usually used written in terms of the stagnation enthalpy.

3-5-3 Physical Properties of Mixtures

The general-purpose computer code of Svehla and McBride (1973) used for most of the physical property data in Appendix A will also calculate properties of mixtures. One should not suppose that a property (e.g., viscosity) of a mixture is a linear function of the concentration between the property values of each single constituent. Some results for mixture viscosity and thermal conductivity for H_2 in air are shown in Fig. 3-4 from Eckert et al. (1958).

For mixtures of only two constituents, the following approximate rules may be used. For the viscosity of gases, the equation of Buddenberg and Wilke (1949) can be used:

$$\mu = \frac{X_1^2}{(X_1^2/\mu_1) + 1.385(X_1 X_2 RT/pW_1 D_{12})}$$

$$+ \frac{X_2^2}{(X_2^2/\mu_2) + 1.385(X_2 X_1 RT/pW_2 D_{21})} \tag{3-66}$$

where X_i is the mole fraction of species i. For the thermal conductivity of gases one can use

$$k = \tfrac{1}{2}\left[(k_1 X_1 + k_2 X_2) + \frac{k_1 k_2}{(X_1\sqrt{k_2} + X_2\sqrt{k_1})^2} \right] \tag{3-67}$$

3-6 MATHEMATICAL OVERVIEW

Before leaving this chapter, which presents the derivation of the exact, boundary layer forms of the equations of motion that will be used throughout the remainder of this book, it is appropriate, indeed essential, to consider the nature of the mathematical problem that has been created. For purposes of discussion, the set of Eqs. (3-27), (3-28), (3-51), and (3-61) with an appropriate equation of state, usually

$$p = \frac{\rho RT}{W} \quad \text{with} \quad W = \left(\sum_i c_i/W_i \right)^{-1} \tag{3-68}$$

can be said to represent the system. To keep these considerations simple, we restrict ourselves to fluids of only one species here.

There are four dependent variables, u, v, ρ, and T [$p(x)$ must be given from the inviscid solution], and four equations. There are three independent variables, x, y, and t, so the system consists primarily [except for Eq. (3-68)] of partial differential equations. The equations also have variable coefficients. More than that, Eqs. (3-27) and (3-51) are nonlinear on the left-hand sides. This can be seen from the existence of terms such as $u\, \partial u/\partial x = \partial(u^2/2)/\partial x$ which clearly have a nonlinear form for the dependent variables. This has the

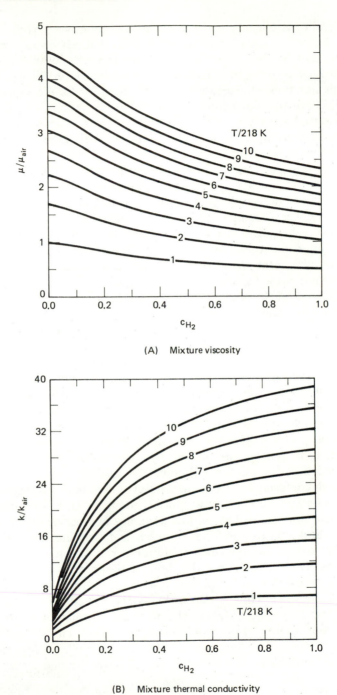

(A)   Mixture viscosity

(B)   Mixture thermal conductivity

Figure 3-4   Variation of mixture viscosity and thermal conductivity for H_2 in air. (From Eckert et al., 1958.)

same general consequences as the occurrence of terms such as yy' or y^2 in an ordinary differential equation for $y(x)$. Methods that rely on superposition for solution cannot be used. This eliminates such powerful techniques as *Fourier series* and *Laplace transforms* from use in boundary layer theory. The exception to this situation is the case of the energy equation for incompressible flow, Eq. (3-53). Then the momentum and continuity equations can be solved for (u, v) uncoupled from the energy equation, since ρ = constant is known *a priori*. The energy equation is rendered linear with known, variable coefficients, and it can be solved taking advantage of that fact.

Let us now turn our attention to the matter of boundary and initial conditions. Looking at the system of equations, one can see that it is second order in y (second derivatives with respect to y are present) and first order in x and t. This implies a requirement for two boundary conditions on y (specified values at two values of y), one boundary condition on x, and one boundary or initial condition on t. The mathematical requirement is matched well with physical intuition. One would expect to impose conditions on the body surface, $y = 0$, and at the edge of the boundary layer, $y = \delta$ or δ_T. Actually, from the point of view of the boundary layer, $y = \delta$ is the same thing as the asymptotic limit $y \to \infty$, since the boundary layer flow will blend asymptotically into the outer inviscid flow. One also would expect to specify the flow at the leading edge or the front stagnation point of a body (i.e., $x = 0$). Finally, for an unsteady flow, the state of the flow at the beginning of the time period of interest, $t = t_0$, clearly must be specified. In an unsteady case, the boundary conditions on x and y can also be arbitrary, but given, functions of time.

The form of the equations and the required boundary conditions produce a natural *front-to-back* (in the streamwise direction) orientation to the mathematical problem. This system is mathematically characterized as *parabolic*, and such systems are solved by *downstream marching* procedures whether analytical or numerical techniques are used.

We risk confusing the reader and observe here that this physically satisfying state of affairs occurs only within the boundary layer approximation for viscous flows. For steady flows at low-Reynolds number or over bluff bodies, the Navier–Stokes equations must be used, and they are not *parabolic*, but *elliptic*. They require two boundary conditions on x. This means that one must not only specify the condition of the flow at the front of the body, which is generally a straightforward task, but also somewhere behind the body, for example in the wake. How is one to know the velocity profile in the wake *before* solving the problem? Obviously, no exact information will be available and some approximate formulation must be employed. This difficulty with the *downstream boundary condition* is one of the major problems now under study in viscous flow research. Those matters are, however, beyond the scope of this book.

PROBLEMS

3.1. For incompressible, constant-property, planar, unsteady flow, write the energy equation using a stream function.

3.2. Consider a flat plate of infinite extent in the x direction. One can then say that $u \neq u(x)$. How does such a condition affect Eqs. (3-6a) and (3-27a)?

3.3. For the same situation as in Prob. 3.2, what is the effect of a porous surface at $y = 0$ with suction [i.e., $v(0, x) = v_w < 0$] or injection ($v_w > 0$)?

3.4. Calculate and plot the variation of the viscosity of a CO_2–air mixture from 0 to 100% at 400 K.

3.5. Calculate and plot the variation of the thermal conductivity of an H_2–air mixture from 0 to 100% at 300 K. Find some exact data in the literature for a point or two and compare with your prediction.

3.6. Consider a release of H_2 from a slit across a uniform N_2 stream at 1 m/s at 400 K and 0.1 atm. The H_2 exits in the same direction and at the same velocity as the N_2 stream. Estimate the rate of spread of the H_2 plume.

4

EXACT AND NUMERICAL SOLUTIONS FOR LAMINAR CONSTANT-PROPERTY INCOMPRESSIBLE FLOWS

4-1 INTRODUCTION

The mathematical formulation of boundary layer problems as developed in Chapter 3 is quite complex, so there is only a small group of cases that admits to an exact analytical solution even if the discussion is restricted to steady, incompressible, constant-property flow. Only a fraction of this small class of problems has any practical application. We will treat the most widely applicable case in the next section.

The next group of exact solutions that exists is for those restrictive classes of flows where helpful simplifications of the equations are rigorously possible. The solutions of the simplified equations must usually be obtained numerically, but simple methods can be used. The third and fourth sections of this chapter deal with some of those flows. These solutions have some practical and scientific value, but they do not apply, for example, to the flow over general bodies of engineering interest.

Books written as recently as a decade ago would have a chapter with this heading consisting of detailed coverage of the material described above together with other exact solutions of even less practical value. However, the current widespread availability of the large digital computer has put *numerically exact* solutions of rather general boundary layer problems within the reach of all engineers and applied scientists. Thus, the main body of this chapter is devoted to numerical methods. Much of that material will also carry over directly into the treatment of turbulent flows.

4-2 FULLY DEVELOPED FLOW IN A TUBE

The flow of a viscous fluid in a round tube or pipe is obviously of great engineering interest. The item of primary practical interest is the loss in pressure head, or *pressure drop*, along the pipe, since that must be made up by pumping. It happens that this flow problem is a useful member of that group where an exact solution of the equations of motion is rather easy to obtain.

Most applications of pipe flows involve cases where the length of the pipe is measured in many, many diameters. In that instance, the details of the flow near the beginning of the pipe are not important. If the entrance to the pipe had sharp lips, a boundary layer similar to that on a flat plate would initially form. However, as the boundary layer grew along the wall, it would interact with the inner inviscid flow and then with that on the opposite wall until finally the flow would be fully viscous. After the boundary layer fills the pipe, it can no longer grow, and an equilibrium state is reached where changes in the boundary layer profile along the pipe cease. This condition is called *fully developed* flow.

The flow problem is shown schematically in Fig. 4-1, and we will analyze the velocity field first. The appropriate equations are the continuity equation, Eq. (3-7a) and the momentum equation, Eq. (3-26), less the unsteady term and assuming constant properties. The fully developed flow condition is stated mathematically as $u = u(r) \neq f(x)$. Substituting into the continuity equation, one sees that $v \equiv 0$ satisfies the equation and the boundary condition, $v(x, R) = 0$. Using this and the previously stated condition on $u(r)$ in the momentum equation produces

$$0 = -\frac{1}{\rho}\frac{dp}{dx} + \frac{v}{r}\frac{d}{dr}\left(r\frac{du}{dr}\right) \tag{4-1}$$

or

$$\frac{v}{r}\frac{d}{dr}\left(r\frac{du}{dr}\right) = \frac{1}{\rho}\frac{dp}{dx} = \text{constant} \tag{4-2}$$

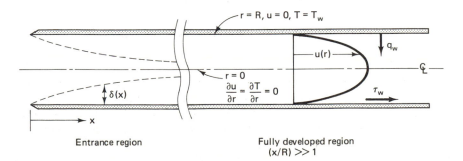

Figure 4-1   Schematic of the flow in a pipe.

This result can be seen from the fact that $u \neq f(x)$ and $p \neq g(r)$. Thus, the equations have been rigorously reduced to a single, *linear*, ordinary differential equation for $u(r)$. This equation can be easily integrated twice to give

$$u(r) = -\frac{1}{4\mu}\frac{dp}{dx}(R^2 - r^2) \tag{4-3}$$

where the boundary conditions

$$u(x, R) = 0$$
$$\frac{du}{dr}(x, 0) = 0 \tag{4-4}$$

have been used. The solution was written with the minus sign in front, since there will be a pressure loss along the tube (i.e., $dp/dx < 0$). The velocity profile described by Eq. (4-3) is a paraboloid, and the geometry of that shape leads to a value of the average velocity across the pipe as

$$u_{\text{ave}} = \left(-\frac{dp}{dx}\right)\frac{R^2}{8\mu} = \frac{u_c}{2} \tag{4-5}$$

The *pressure drop* is caused by the viscous shear on the pipe wall, and that relation can be developed using Eq. (4-3) as

$$\tau_w = -\mu\left.\frac{du}{dr}\right|_{r=R} = -\frac{R}{2}\frac{dp}{dx} \tag{4-6}$$

The *pressure drop* per unit length along the pipe is usually expressed in terms of a *resistance coefficient*, λ, defined as

$$\lambda \equiv \frac{(-dp/dx)D}{\frac{1}{2}\rho u_{\text{ave}}^2} \tag{4-7}$$

Substituting for dp/dx from Eq. (4-5) yields

$$\lambda = \frac{64}{\rho u_{\text{ave}}(2R)/\mu} = \frac{64}{\text{Re}_D} \tag{4-7a}$$

Note two additional facts about this solution. First, experiment indicates that flow in a pipe remains laminar up to about $\text{Re}_D = 2300$. Second, Eq. (4-1) is actually the correct, reduced form of the full equations of motion for this problem, and the boundary layer assumption was not necessary in order to obtain this solution.

4-3 SIMILAR SOLUTIONS

There is a second class of flow problems that admits to relatively simple, exact solutions. Again, the conditions are such that simplifications can be rigorously made in the equations of motion. This class of problems is somewhat more general than that typified by fully developed pipe flow, but the simplifications of the equations are not as great, so that the solutions are harder to obtain.

If one considers steady, planar, constant-density, constant-property flows, the mathematical problem is posed by Eqs. (3-6a) and (3-27a). The difficulty is that these are nonlinear, partial differential equations. For the class of problems in the preceding section, the terms on the left-hand side of the momentum equation disappear, and the system becomes linear. Those problems are specifically chosen such that this happens, but that cannot be expected to occur for general problems.

On the other hand, is it possible to find a class of problems such that the partial differential equations reduce rigorously to ordinary differential equations, even if nonlinear? The reader has seen a comparable maneuver before with the method of *separation of variables*. There, one hopes to find cases where

$$u(x, y) = X(x)Y(y) \tag{4-8}$$

giving two ordinary differential equations rather than one partial differential equation. Unfortunately, that procedure will not work for boundary layer problems. What then? Certainly, problems where the x or y dependence would disappear altogether cannot be very general. The last possibility is the introduction of one, new independent variable η that is a function of (x, y) such that

$$u(x, y) = u[\eta(x, y)] \tag{4-9}$$

Of course, not only the equations but the boundary conditions must be compatible with this representation. That surely will not be possible for all flow problems, but perhaps there are some of this type.

The development of these solutions proceeded logically from two roots. First, it had been known for some time from experiment that flat-plate boundary layer profiles remained unchanged along the plate if plotted as u/U_e versus y/δ. We say that each normalized profile is *similar* to every other. Further, experiment showed that $\delta \sim \sqrt{x}$, so profiles plotted as u/U_e versus $y/\sqrt{x}$ would also be *similar*.

The second root led from solutions to related problems for unsteady, one-dimensional heat conduction in solids. Such problems are governed by

$$\frac{\partial T}{\partial t} = \frac{k}{\rho c_p} \frac{\partial^2 T}{\partial y^2} \tag{4-10}$$

and the book by Carslaw and Jaegar (1959) is an encyclopedia of solutions to different problems. This is a *linear, parabolic*, partial differential equation, so it is much easier to solve than our boundary layer problem. Nonetheless, it can serve well as a simple *model equation* of the general *parabolic* type. One finds that the grouping $y/\sqrt{t}$ occurs very frequently in solutions. A crude approximation heuristically, not rigorously, made to Eq. (3-27a) for steady flow over a flat plate, $dp/dx \equiv 0$, would be

$$U_e \frac{\partial u}{\partial x} \approx v \frac{\partial^2 u}{\partial y^2} \qquad (4\text{-}11)$$

The streamwise variable can be combined with U_e to produce a pseudotime, x/U_e. Solutions to problems governed by this equation would then often involve $y/\sqrt{x}$, since solutions to Eq. (4-10) often involve $y/\sqrt{t}$. This line of reasoning is not rigorous, but it also strongly suggests that a promising trial form for $\eta(x, y)$ would be $\eta \sim y/\sqrt{x}$.

4-3-1   Exact Solution for Flow over a Flat Plate

This solution was originally produced by Blasius (1908), who was a doctoral student under Prandtl. It is convenient to use the stream function formulation, since then only one equation for one unknown $\psi(x, y)$ is involved. That equation is, from Eq. (3-29) for steady flow with $dp/dx = 0$,

$$\frac{d\psi}{\partial y} \frac{\partial^2 \psi}{\partial y \, dx} - \frac{\partial \psi}{\partial x} \frac{\partial^2 \psi}{\partial y^2} = v \frac{\partial^3 \psi}{\partial y^3} \qquad (4\text{-}12)$$

We choose to write the assumption in Eq. (4-9) as

$$\frac{u}{U_e} = Af'(\eta) \qquad (4\text{-}13)$$

with, $\eta = By/\sqrt{x}$, where A and B are constants to be determined later to simplify the final equation. Now, it cannot be expected that $\psi = \psi(\eta)$ alone, since the value of ψ at any point y above the plate determines the total mass flow between $y = 0$ and the given y. We know that the boundary layer grows proportional to $\sqrt{x}$, so it can be conjectured that

$$\psi \sim \sqrt{x} f(\eta) = C\sqrt{x} f(\eta) \qquad (4\text{-}14)$$

where C is another *convenience* constant. The form of $f(\eta)$ in Eqs. (4-13) and (4-14) can be seen as correct by reexamining the definition of the stream function, Eq. (3-8), and noting that $\eta \sim (y/\sqrt{x})$. Substituting into Eq. (4-12) and collecting terms, one finds it convenient to select A, B, and C so that

$$\eta = \tfrac{1}{2} y \sqrt{\frac{U_e}{vx}}$$

$$\frac{u}{U_e} = \tfrac{1}{2} f'(\eta)$$

$$\psi = \sqrt{vU_e x} f(\eta) \qquad (4\text{-}15)$$

$$v = \frac{1}{2} \sqrt{\frac{vU_e}{x}} (\eta f' - f)$$

giving the final equation

$$f''' + ff'' = 0 \tag{4-16}$$

which surely has a simple appearance, although it is important to note the remaining nonlinearity in the term ff''.

This looks like good progress, but the boundary conditions must still be examined. Indeed, it is the boundary conditions that determine which problems admit to such a solution and which do not. That is also the case with *separation of variables*. Begin with the usual boundary conditions written in terms of (u, v) and (x, y):

$$y = 0, \ x \geq 0: \quad u(x, 0) = v(x, 0) = \psi(x, 0) = 0$$

$$y \rightarrow \infty, \ \text{all} \ x: \quad u(x, y) \rightarrow U_e \tag{4-17}$$

$$x = 0, \ y > 0: \quad u(0, y) = U_e$$

The last condition imposes a uniform approach stream with velocity U_e. These conditions must be cast in terms of $\eta \sim y/\sqrt{x}$. Note that η cannot tell the difference between $y \rightarrow \infty$ or $x \rightarrow 0$ with $y > 0$. This is accommodated in Eq. (4-17), but it is worth observing that a uniform approach flow is *required*. Not all flat-plate boundary layer problems have similar solutions. Finally, the boundary conditions of interest here can be written in terms of η as:

$$\eta = 0: \quad f(0) = f'(0) = 0$$
$$\eta \rightarrow \infty: \quad f'(\eta) \rightarrow 2.0 \tag{4-18}$$

Blasius solved Eq. (4-16) with (4-18) using inner and outer series expansions that are joined within the boundary layer. A simpler method was introduced later by Piercy and Preston (1936), based on the notion of successive approximations. It is well suited to the computer. Write $f''(\eta) = z$, and Eq. (4-16) becomes

$$\frac{dz}{d\eta} = -fz \tag{4-19}$$

Consider $f(\eta)$ as known; then

$$\frac{dz}{z} = -f \, d\eta \tag{4-19a}$$

which can be integrated to give

$$\ln(z) = -\int f \, d\eta + \ln(C_1) \tag{4-19b}$$

or

$$z = C_1 e^{-\int f \, d\eta} \equiv f'' \tag{4-19c}$$

This can be integrated to give f' or u/U_e through Eq. (4-15):

$$\frac{u}{U_e} = \tfrac{1}{2} f' = \frac{1}{2} \int z \, d\eta + C_2 = \frac{C_1}{2} \int e^{-\int f \, d\eta} \, d\eta + C_2 \qquad (4\text{-}20)$$

When C_1 and C_2 are determined from the boundary conditions, there results

$$\frac{u}{U_e} = \frac{\displaystyle\int_0^\eta e^{-\int_0^\eta f \, d\eta} \, d\eta}{\displaystyle\int_0^\infty e^{-\int_0^\eta f \, d\eta} \, d\eta} \qquad (4\text{-}21)$$

The procedure is as follows. First, guess a profile for $f(\eta)$ (e.g., a straight line, $f = 2\eta$) and introduce that into Eq. (4-21), which yields an approximation to u/U_e. Integrate that once to obtain an updated $f(\eta)$, which can be substituted into Eq. (4-21) to get a better approximation, and so on. Repeat the process until convergence. Few iterations are required. Another method suited to the computer is called *shooting*. The real complication here is that the boundary conditions are given at two points [see Eq. (4-18)]. A straightforward, third-order problem would have $f(0), f'(0),$ and $f''(0)$ given. One can, however, treat the problem at hand in a similar fashion by guessing $f''(0)$ and calculating out to large η to see if $f'(\eta) \rightarrow 2.0$. If not, a new guess on $f''(0)$ is made and the procedure is repeated until convergence.

The solution is shown in Fig. 4-2 and Table 4-1. It is common to take the outer edge of the boundary layer as corresponding to $\eta = 2.6$ (i.e., $u/U_e = 0.994$). From Eqs. (4-15) we then obtain

$$\delta(x) = \frac{5.2x}{\sqrt{\text{Re}_x}} \qquad (4\text{-}22)$$

Using the profiles in Table 4-1 in Eqs. (2-12) and (2-13) gives

$$\delta^* = \frac{1.721x}{\sqrt{\text{Re}_x}} \qquad (4\text{-}23)$$

and

$$\theta = \frac{0.664x}{\sqrt{\text{Re}_x}} \qquad (4\text{-}23a)$$

Finally, the solution gives

$$C_f = \frac{f''(0)}{2\sqrt{\text{Re}_x}} = \frac{0.664}{\sqrt{\text{Re}_x}} \qquad (4\text{-}24)$$

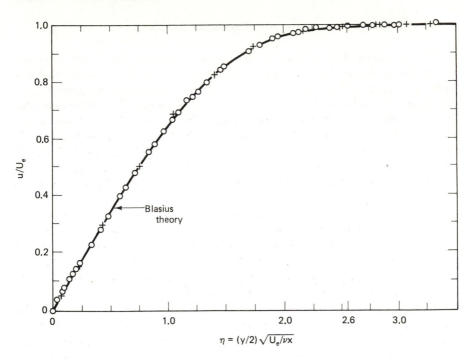

Figure 4-2 Comparison between Blasius solution and experiments of Nikuradse for laminar flow over a flat plate. (From Nikuradse, 1942.)

TABLE 4-1   Tabulated Blasius Solution of the
Flat-Plate Problem

η	$f(\eta)$	$f'(\eta)$	η	$f(\eta)$	$f'(\eta)$
0.0	0.0	0.0	1.4	1.2310	1.6230
0.1	0.0066	0.1328	1.5	1.3968	1.6921
0.2	0.0266	0.2655	1.6	1.5691	1.7522
0.3	0.0597	0.3979	1.7	1.7470	1.8035
0.4	0.1061	0.5294	1.8	1.9295	1.8467
0.5	0.1656	0.6596	1.9	2.1161	1.8822
0.6	0.2380	0.7876	2.0	2.3058	1.9110
0.7	0.3230	0.9125	2.2	2.6924	1.9517
0.8	0.4203	1.0335	2.4	3.0853	1.9756
0.9	0.5296	1.1495	2.6	3.4818	1.9885
1.0	0.6500	1.2595	2.8	3.8803	1.9950
1.1	0.7812	1.3626	3.0	4.2796	1.9980
1.2	0.9223	1.4580	3.5	4.2793	1.9998
1.3	1.0725	1.5449	4.0	6.2792	2.0000

The figure shows u/U_e on the vertical axis plotted against $\eta = (y/2)\sqrt{U_e/\nu x}$ on the horizontal axis, with the Blasius theory curve labeled.

4-3-2 Similar Solutions with Pressure Gradient

It was stated before that the similarity technique would not work for general body shapes [i.e., essentially arbitrary $U_e(x)$], and one should not expect it to on purely physical grounds. However, there are some special cases with pressure gradients, $U_e(x) \neq$ constant, that do admit to solutions of this type. By substituting a fairly general assumed form into the momentum equation, collecting terms, and then insisting that all explicit dependence on x alone disappear, it emerges that cases where $U_e(x) = U_1 x^m$ will serve. This is a special case, and it corresponds to the inviscid flow over a wedge with an opening angle, $\beta\pi = 2m\pi/(m + 1)$. The momentum equation takes the form

$$f''' + ff'' - \frac{2m}{m+1}\,[(f')^2 - 1] = 0 \qquad (4\text{-}25)$$

with

$$\eta_1 = y\sqrt{\frac{m+1}{2}\,\frac{U_e}{\nu x}}$$

$$\psi = \sqrt{\frac{2}{m+1}}\,\sqrt{\nu U_1 x^{m+1}}\,f(\eta_1) \qquad (4\text{-}26)$$

This case was derived by Falkner and Skan (1930), and the solutions shown in Fig. 4-3 were obtained by Hartree (1937). The case $\beta = m = 0$ is the

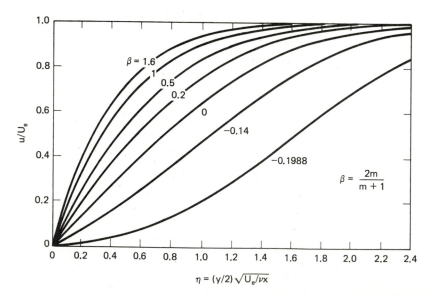

Figure 4-3 Velocity profiles for laminar flow over wedges of various angles. (From Hartree, 1937.)

TABLE 4-2   **Solutions of the Falkner–Skan Wedge Flows**

m	β	$C_f \sqrt{Re_x}$	$\dfrac{\delta^*}{x}\sqrt{Re_x}$	$\dfrac{\theta}{x}\sqrt{Re_x}$
1	1	2.465	0.648	0.292
$\frac{1}{3}$	0.5	1.515	0.985	0.429
0.1	0.182	0.993	1.348	0.557
0	0	0.664	1.721	0.664
-0.01	-0.0202	0.623	1.780	0.679
-0.05	-0.105	0.427	2.117	0.751
-0.0904	-0.1988	0	3.428	0.868

flat-plate problem. Values of $\beta < 0$ represent *adverse* pressure gradients, and $\beta = -0.1988$ corresponds to separation (see Fig. 4-3). Cases with $\beta > 0$ represent favorable gradients, and $\beta = m = 1$ is the case of planar flow at a stagnation point. Some of the gross features of the solutions are tabulated in Table 4-2.

It is important to remember here that a similar solution means that all profiles for any given case have the same nondimensional shape. Thus, the separation profile for $\beta = -0.1988$ does not develop at some axial station along the body. For that case, the profiles at *every* axial station along the body are separated. These comments should help the reader grasp just how *special*, and not general, these cases really are.

4-4 SOLUTIONS TO THE LOW-SPEED ENERGY EQUATION

4-4-1 Fully Developed Flow in a Tube

Many devices employ pipes or tubes to promote heat transfer (e.g., heat exchangers) and other pipe flow cases simply involve heat transfer because a temperature difference exists between the wall and the fluid. If the temperature difference is not too large, the problem can still be analyzed retaining the incompressible, constant-property assumptions. The appropriate energy equation is the low-speed, steady form of Eq. (3-55).† Using the fully developed flow conditions, the equation rigorously reduces to

$$\rho c_p u \frac{\partial T}{\partial x} = \frac{k}{r}\frac{\partial}{\partial r}\left(r\frac{\partial T}{\partial r}\right) \tag{4-27}$$

where $u(r)$ is known from Eq. (4-3).

---

† Note that there is no difference between c_p and c_v for the special case of a constant-density, constant-property fluid.

The boundary conditions on $T(r)$ are

$$\left.\frac{\partial T}{\partial r}\right|_{r=0} = 0$$

$$T(x, R) = T_w(x) \quad \text{or} \quad -\left(-k\left.\frac{\partial T}{\partial r}\right|_{r=R}\right) = q_w(x) \tag{4-28}$$

That is, either the wall temperature or the wall heat transfer rate may be specified. A simple case occurs with the use of a specified constant q_w. Consider an energy balance for the fluid flowing through two adjacent stations along the tube under that condition. It is clear that the average value of the temperature must increase linearly with distance x along the tube. For *fully developed* flow, this must also hold at any value of the radius r, so we can say that $\partial T/\partial x = $ constant. Equation (4-27) can be easily integrated for $\partial T/\partial x = $ constant to give

$$T(r) - T_w = \frac{2u_{ave}R^2}{(k/\rho c_p)}\left(\frac{\partial T}{\partial x}\right)\left[\frac{1}{4}\left(\frac{r}{R}\right)^2 - \frac{1}{16}\left(\frac{r}{R}\right)^4 - \frac{3}{16}\right] \tag{4-29}$$

after the boundary conditions are applied. It is common to introduce the average or *bulk* temperature

$$T_b \equiv \frac{\displaystyle\int_0^R \rho u c_p T 2\pi r \, dr}{\displaystyle\int_0^R \rho u c_p 2\pi r \, dr}$$

$$= T_w - \frac{11}{48}\frac{u_{ave}R^2}{k/\rho c_p}\left(\frac{\partial T}{\partial x}\right) \tag{4-30}$$

Basing a film coefficient on the difference between the wall and bulk temperatures,

$$q_w = h(T_w - T_b) \tag{4-31}$$

we get

$$\text{Nu}_D \equiv \frac{h(2R)}{k} = 4.364 \tag{4-32}$$

For constant T_w this value is 3.65.

4-4-2 Similar Solutions

For steady, low-speed, constant-density, constant-property flows, the exact energy equation becomes

$$u\frac{\partial T}{\partial x} + v\frac{\partial T}{\partial y} = \frac{\nu}{\text{Pr}}\frac{\partial^2 T}{\partial y^2} \tag{4-33}$$

Comparing this to the momentum equation for flow over a flat plate, Eq. (3-27a), (without the first terms on both the left- and right-hand sides), one can anticipate that there will be some situations where *similar* solutions also exist for the energy equation.

For a flat plate with a constant wall temperature and a uniform temperature approach flow, the boundary conditions are

$$y = 0, \ x \geq 0: \quad T = T_w = \text{constant}$$

$$y \rightarrow \infty, \ \text{all} \ x: \quad T = T_e = \text{constant} \qquad (4\text{-}34)$$

$$x = 0, \ y > 0: \quad T = T_e = \text{constant}$$

The temperature is more conveniently written in terms of a so-called *excess temperature*

$$\Theta(x, y) \equiv \frac{T - T_w}{T_e - T_w} \qquad (4\text{-}35)$$

which ranges from zero to unity. Using this and the definitions in Eq. (4-15), Eq. (4-33) becomes

$$\Theta'' + \text{Pr} \cdot f \Theta' = 0 \qquad (4\text{-}36)$$

This equation is linear because $f(\eta)$ is known from the prior solution of the momentum equation, and it is simply a variable coefficient here.

The solution to Eq. (4-36) can be obtained by a similar procedure as for the momentum equation

$$\Theta = \frac{\displaystyle\int_0^{\eta} e^{-\int_0^{\eta} \text{Pr} \cdot f \, d\eta} \, d\eta}{\displaystyle\int_0^{\infty} e^{-\int_0^{\eta} \text{Pr} \cdot f \, d\eta} \, d\eta} \qquad (4\text{-}37)$$

Temperature profiles plotted as

$$1 - \Theta = \frac{T - T_e}{T_w - T_e}$$

are given in Fig. 4-4. They show, when compared with the vel corresponding to a flat plate in Fig. 4-2 ($\beta = 0$), an important Prandtl number. For $\text{Pr} = 1.0$, the nondimensional profiles are $\text{Pr} > 1$, the temperature profiles are *fuller*, and $\delta_T < \delta$. For Pr situation holds (remember that $\text{Pr} = 0.7$ for air and most ga' the heat transfer coefficient is

$$\text{Nu}_x \equiv \frac{hx}{k} = 0.332 \text{Pr}^{1/3} \sqrt{\text{Re}_x}$$

There are also similar solutions to Eq. (4-33) c' flows, $U_e \sim x^m$, for the momentum equation. The'

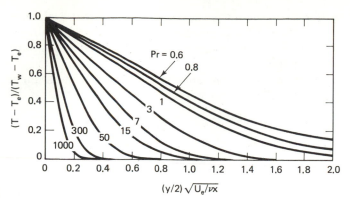

Figure 4-4 Temperature profiles for laminar flow over a constant-temperature flat plate. (From Eckert and Drewitz, 1940.)

Falkner (1931) for $U_e \sim x^m$ and/or $(T_w - T_e) \sim x^n$. The energy equation for these cases is

$$\Theta'' + \mathrm{Pr} \cdot f\Theta' - 2\mathrm{Pr} \cdot nf'(\Theta - 1) = 0 \qquad (4\text{-}40)$$

Some solutions are presented in Figs. 4-5 and 4-6. The profiles in Fig. 4-5 show the effect of pressure gradient ($m \neq 0$) and Prandtl number for the case of a constant wall temperature ($n = 0$), and those in Fig. 4-6 show the effect of

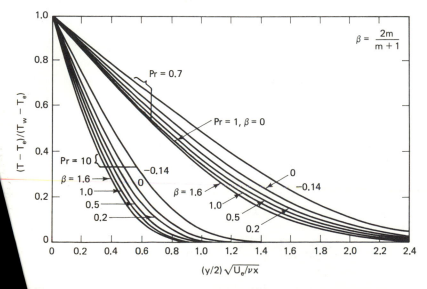

Figure 4-5 Temperature profiles for laminar flow over constant-temperature edges. (From Eckert, 1942.)

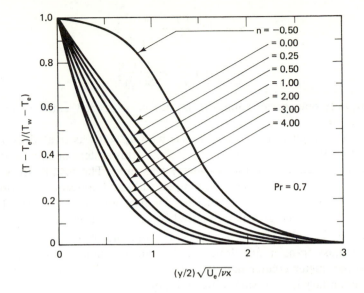

Figure 4-6   Temperature profiles for laminar flow over variable-temperature flat plates. (From Levy, 1952.)

variable wall temperature ($n \neq 0$) for a flat plate ($m = 0$) at $Pr = 0.7$. Comparing the two, one can note that the effect of a variable wall temperature is larger than that of a pressure gradient.

4-5 FOREIGN FLUID INJECTION

4-5-1 Similar Solutions

Under the assumptions of constant density and properties taken for this chapter, only rather restricted cases of foreign gas injection can be treated. More general cases are discussed in Chapter 5. Here, we must assume that not only are the properties independent of temperature and pressure but that the properties of the two fluids are essentially equal. Taking this all together, the complete system of equations to be used in such a case for a steady flow over a flat plate is

$$\frac{\partial u}{\partial x} + \frac{\partial v}{\partial y} = 0$$

$$u \frac{\partial u}{\partial x} + v \frac{\partial u}{\partial y} = v \frac{\partial^2 u}{\partial y^2}$$

$$(4\text{-}41)$$

$$u \frac{\partial T}{\partial x} + v \frac{\partial T}{\partial y} = \frac{\nu}{Pr} \frac{\partial^2 T}{\partial y^2}$$

$$u \frac{\partial c_i}{\partial x} + v \frac{\partial c_i}{\partial y} = \frac{\nu}{Sc} \frac{\partial^2 c_i}{\partial y^2}$$

Actually, only one species equation, say for c_1, is needed, since we know that $c_1 + c_2 \equiv 1.0$.

A problem governed by Eqs. (4-41) with $v_w \neq 0$ was treated by Hartnett and Eckert (1957). In a case where $v_w \sim x^{-1/2}$, this system of equations admits to a similar solution, and that property was used. The results are shown in Figs. 4-7, 4-8, and 4-9 for velocity, temperature, and concentration profiles in the boundary layer and film coefficients for heat and mass transfer. The profiles can be seen to be greatly influenced by either suction $[(v_w/U_e)\sqrt{Re_x} < 0]$ or injection $[(v_w/U_e)\sqrt{Re_x} > 0]$. For injection, the wall friction and surface heat and mass transfer are dramatically reduced. Indeed, for values of the injection parameter greater than 0.62, the surface transfer rates all go to zero, and the boundary layer is said to be *blown off*.

The transverse mass flux of the air (species 2) at any point is composed of a part due to a *diffusive velocity* and a part due to the *convective velocity v*. Thus,

$$\rho_2 v_2 = -\rho D_{12} \frac{\partial c_2}{\partial y} + \rho c_2 v \qquad (4\text{-}42)$$

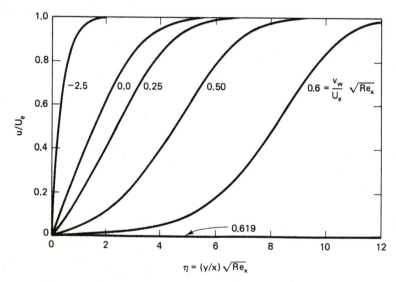

Figure 4-7 Velocity profiles for fluid injection from a flat plate. (From Hartnett and Eckert, 1957.)

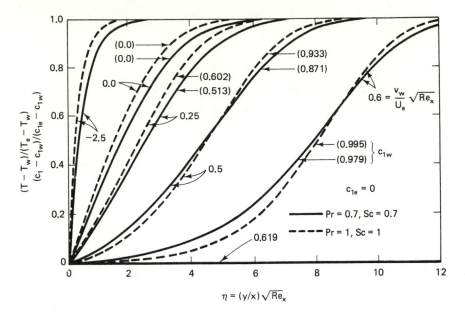

Figure 4-8   Temperature and mass fraction profiles for fluid injection from a flat plate. (From Hartnett and Eckert, 1957.)

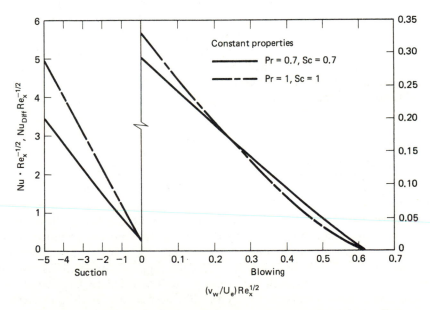

Figure 4-9   Nondimensional heat and mass transfer rates for fluid injection from a flat plate. (From Hartnett and Eckert, 1957.)

or

$$v_2 = -\frac{D_{12}}{c_2}\frac{\partial c_2}{\partial y} + v \qquad (4\text{-}42a)$$

Since no air penetrates into the wall, $v_2(x, 0) \equiv 0$, and

$$v_w = \frac{D_{12}}{c_{2w}}\frac{\partial c_2}{\partial y}\bigg|_w \qquad (4\text{-}42b)$$

Using $c_1 + c_2 \equiv 1.0$, this can be rewritten

$$v_w = -\frac{D_{12}}{1 - c_{1w}}\frac{\partial c_1}{\partial y}\bigg|_w \qquad (4\text{-}42c)$$

Therefore, one is not free to prescribe both v_w and c_{1w}.

4-5-2 Fully Developed Flow in a Tube

It is a simple matter to envision a fully developed flow in a tube or a pipe with mass transfer. The wall might, for example, be covered with a material that is soluble in the fluid flowing in the pipe. That case is analogous to a constant-temperature wall. On the other hand, the walls might be porous with a foreign fluid forced in at a set rate. That case is analogous to a constant wall heat flux.

In Chapter 2 we observed a close interrelationship among momentum, heat, and mass transfer. Thus, the results for heat transfer in a pipe presented in Sec. 4-4-1 can be used here by replacing the Nusselt number with the Nusselt number for diffusion and the Prandtl number with the Schmidt number *if* the transverse velocity at the wall is small and its influence can be neglected. A better procedure would be to begin with results for heat transfer in the presence of injection or suction (with the flow still restricted to one species) and then apply the analogy between heat and mass transfer.

4-6 NUMERICALLY EXACT SOLUTIONS

Up to this point, the reader has seen representative examples of all the classes of exact solutions of the boundary layer equations for low-speed flows. Clearly, the scope of problems covered by these solutions is extremely limited, and they do not cover cases of such fundamental interest to the engineer as flow over a ship hull or through a pump. During the period between the development of the boundary layer theory in 1904 and the 1960s, the practicing engineer had little use for exact solutions and had to rely on approximate methods, usually based on the integral momentum equation as described in Chapter 2. Beginning about 1960, engineers began to gain access to digital computers of sufficient size and speed to undertake essentially exact solutions to more general flow problems. It is interesting to note that these machines had generally been purchased for bookkeeping, and engineers and scientists were only allowed to use them on nights and weekends in the beginning of the computer age.

Today, one often hears or reads the statement that some complex differential equation was *solved* with a digital computer. Actually, such a statement can be misleading. Only recently have computer routines for symbolic operations in algebra and calculus been developed.† These routines can only do the equivalent of that which mathematicians can do by hand. Since there are no general methods for exactly solving nonlinear partial differential equations by hand, the computer cannot do so either. The real power of the digital computer is that it can do *arithmetic* very rapidly on a very large scale, without errors. But boundary layer problems are not arithmetic problems, they are differential equation problems, so how can the computer be used to solve them? The answer is simple; one must reduce the solution of the differential equations to an essentially arithmetic problem. The only way to do that involves undoing one of the great intellectual developments in all of human thought—the *limit process*. The reader can recall his or her introduction to the calculus, involving the concept

$$\frac{dF}{dx}\bigg|_{x=x_0} \equiv \lim_{\Delta x \to 0} \frac{F(x_0 + \Delta x) - F(x_0)}{\Delta x} \tag{4-43}$$

The extension of this idea leads to differential equations. If the process is reversed, a differential equation becomes a series of algebraic relations between the values of the dependent variables at various specific values of the independent variables separated by small *finite differences* (e.g., Δx and Δy).

Consider the situation shown in Fig. 4-10. The region of interest in the flow is overlaid with a grid with small, but finite spacing $(\Delta x, \Delta y)$. We will be concerned with the values of the dependent variables at the intersections of the grid lines, called *node* points. Clearly, there will be many such points, and a simple, unambiguous notation is required to identify them. It is convenient to use an integer pair (n, m) for this purpose. The index n identifies location in the x direction. The initial station, $x = x_i$, is denoted by $n = 1$, and n runs up to a value of $n = N$ at the last station, $x = x_f$. For the transverse coordinate, pick $m = 1$ at $y = 0$ and $m = M$ at $y = y_{max}$, where y_{max} must clearly be greater than $\delta(x)$. Thus,

$$x = x_i + (n - 1)\,\Delta x \tag{4-44a}$$

$$y = (m - 1)\,\Delta y \tag{4-44b}$$

A pair of specific integer values (n, m) then locates a specific node point in the region of interest and a dependent variable is written as $u(x, y) \to u_{n,m}$.

Before proceeding, it should be noted that the analyst must pick the values of (N, M) for a given problem. This is the same as picking $(\Delta x, \Delta y)$.

† The article "Computer Algebra" by R. Pavelle, M. Rothstein, and J. Fitch in *Scientific American*, December 1981, provides interesting background information.

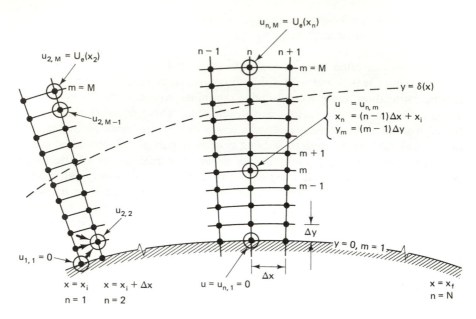

Figure 4-10 Schematic of the grid system and notation for the finite difference solution of boundary layer problems.

This choice must be based on a good estimate of the rapidity of variation expected in the dependent variables with (x, y). Looking at Fig. 4-2, for example, one might say that such a simple velocity profile could be adequately described by 10 to 20 points across the layer. The boundary layer theory tells us that $\partial u/\partial y \gg \partial u/\partial x$, so we would certainly expect to be able to use a $\Delta x > \Delta y$. The choice of $(\Delta x, \Delta y)$ is also strongly influenced by the boundary conditions for a given problem. If a problem involves a rapidly varying $U_e(x)$, a smaller Δx will be required. All of this may seem rather nonspecific, but that is the true state of affairs. The analyst must make these choices; the computer, in general, cannot. Only experience can make this task easier, but most general-purpose computer codes have helpful suggestions for the novice in the user's manual. Finally, it is always prudent to run a sample calculation with a smaller grid size than seems required to check the dependence of the solution on grid size. One is always presuming that the finite differences chosen are small enough that the numerical solution obtained approaches the solution to the differential equation which is based on the limit as $(\Delta x, \Delta y) \to 0$. This whole business is of more than abstract interest. Indeed, it could hardly be of more practical concern, since it involves money—sometimes a lot of it. Obviously, the smaller the choices for $(\Delta x, \Delta y)$, the bigger and more expensive will be the calculations. From the practical side, then, any $(\Delta x, \Delta y)$ smaller than really required to approximate closely the exact solution to the differential equations is too small.

Let us return now to manipulating the differential equations so that the computer can handle them. In principle, the procedure is simple. Consider a first partial derivative

$$\frac{\partial u}{\partial y}\bigg|_{x_n,\,y_m} \approx \frac{\Delta u}{\Delta y} = \frac{u_{n,\,m+1} - u_{n,\,m}}{\Delta y} \tag{4-45a}$$

$$= \frac{u_{n,\,m} - u_{n,\,m-1}}{\Delta y} \tag{4-45b}$$

$$= \frac{u_{n,\,m+1} - u_{n,\,m-1}}{2(\Delta y)} \tag{4-45c}$$

The first of these is easy to understand. One is just taking the difference between the value of the dependent variable at a point and that for a slightly greater value of the independent variable and then dividing by the spacing. This is the same as Eq. (4-43) without the limit process: hence $\approx$, not $=$. A moment's reflection will reveal that the second expression involves the same ideas. Equation (4-45a) is called a *forward* difference, and Eq. (4-45b) is called a *backward* difference. In the limit $\Delta y \to 0$, they are the same thing, but here the limit process is not involved, and they are not the same. They represent two, separate approximations to $\partial u/\partial y$. Each of these approximations has a *one-sided* nature, so the *central* difference scheme in Eq. (4-45c) is preferred in some instances.

An approximation for the second partial derivative follows directly along the same line of thought:

$$\frac{\partial^2 u}{\partial y^2}\bigg|_{x_n,\,y_m} = \frac{\partial}{\partial y}\left(\frac{\partial u}{\partial y}\right)\bigg|_{x_n,\,y_m} \approx \frac{\Delta(\Delta u)}{(\Delta y)^2}$$

$$\approx \frac{(u_{n,\,m+1} - u_{n,\,m}) - (u_{n,\,m} - u_{n,\,m-1})}{(\Delta y)^2}$$

$$\approx \frac{u_{n,\,m+1} - 2u_{n,\,m} + u_{n,\,m-1}}{(\Delta y)^2} \tag{4-46}$$

Even-order derivatives come out directly as *central* differences.

It may look now as if we are ready to proceed and solve all possible boundary layer problems. All that appears necessary is to insert approximations such as Eqs. (4-45) and (4-46) into the boundary layer equations and develop a computer code to perform the necessary calculations. But two important matters remain. The first is an analysis of the behavior of the solution of the finite difference equations compared to the behavior of the solution of the original differential equations and also the behavior of the finite difference solutions themselves. Unfortunately, this analysis cannot be rigorously performed for a nonlinear system. One is forced to analyze a related, linear problem and carry over the results by *analogy*. This is obviously perilous, and a conservative attitude is necessary. This matter will be discussed in the next section. The second matter not yet discussed is boundary and initial conditions. Perhaps surprisingly, that is a simple item that emerges naturally as part of the discussion.

4-6-1 Numerical Analysis of the Linear, Model Equation

The linear, *model*, parabolic equation was introduced earlier, Eq. (4-11) [or its antecedent Eq. (4-10)]. Equation (4-10) is known as the *heat equation*, and its numerical solution has been thoroughly analyzed. The discussion here follows that in the text by Carslaw and Jaegar (1959) adapted to Eq. (4-11).

For the term $\partial u/\partial x$ in Eq. (4-11), it appears that three choices following from Eq. (4-45a), (4-45b), or (4-45c) are available. Actually, that is not so. For this parabolic system, the method of solution is to start from a given initial profile, $u(x_i, y)$, and *march* downstream. At $x = x_i$, the approximation for $\partial u/\partial x$ corresponding to Eq. (4-45b) would involve information at $x = x_i - \Delta x$, which is unavailable. The same thing holds for Eq. (4-45c). Clearly, then, a *forward* difference must be used:

$$\left.\frac{\partial u}{\partial x}\right|_{x_n, y_m} \approx \frac{u_{n+1, m} - u_{n, m}}{\Delta x} \tag{4-47}$$

One could arrive at the same conclusion by arguing that it is necessary to step forward from the initial station in order to begin to march downstream. Only an expression such as Eq. (4-47) will involve information at a downstream location.

Substituting Eqs. (4-47) and (4-46) into Eq. (4-11) gives

$$U_e \frac{(u_{n+1, m} - u_{n, m})}{\Delta x} = v \frac{(u_{n, m+1} - 2u_{n, m} + u_{n, m-1})}{(\Delta y)^2} \tag{4-48}$$

This can be rearranged into an *explicit* relation for $u_{n+1, m}$, the velocity in the boundary layer at a downstream station, in terms of information at the initial station,

$$u_{n+1, m} = Q(u_{n, m+1} + u_{n, m-1}) - (2Q - 1)u_{n, m} \tag{4-48a}$$

where $Q \equiv v(\Delta x)/(U_e(\Delta y)^2)$ involves only the parameters of the problem and the step sizes. This is called an *explicit finite difference approximation*.

By referring to Fig. 4-10, the method of solution can be described. We begin at $x = x_i$ with a known initial profile written here as $u_{1, m}$ for all $1 \leq m \leq M$. The boundary conditions are $u = 0$ on the wall (or $u_{n, 1} = 0$) and $u = U_e$ for large y (or $u_{n, M} = U_e$). The velocity at $x = x_i + \Delta x$ and $y = \Delta y$, the first interior point on the new profile, can be found using Eq. (4-48a) as

$$u_{2, 2} = Q(u_{1, 3} + \overset{0}{u_{1, 1}}) - (2Q - 1)u_{1, 2} \tag{4-49a}$$

where $u_{1, 1} = 0$, since it is on the wall, $m = 1$. The values $u_{1, 3}$ and $u_{1, 2}$ are

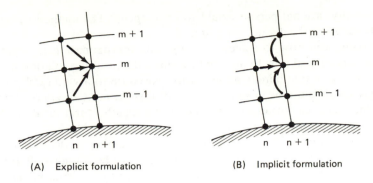

(A) Explicit formulation   (B) Implicit formulation

Figure 4-11 Schematic illustration of the flow of information in marching down-stream for the explicit and implicit finite difference formulations.

known since they are on the given initial profile. Notice how information at three points on the known profile feeds the value at one point on the new profile (see Fig. 4-11(A)).

The velocity at the next point up on the profile, $x = x_i + \Delta x$ and $y = 2(\Delta y)$ can be found from

$$u_{2,3} = Q(u_{1,4} + u_{1,2}) - (2Q - 1)u_{1,3} \qquad (4\text{-}49\text{b})$$

Here, $u_{1,4}$, $u_{1,2}$, and $u_{1,3}$ are all known, since they are on the initial profile. The process is continued up to the next to the last y point, $x = x_i + \Delta x$, $y = (M - 1)(\Delta y)$, where we have

$$u_{2,M-1} = Q(u_{1,M} + u_{1,M-2}) - (2Q - 1)u_{1,M-1} \qquad (4\text{-}49\text{c})$$

Note, $u_{1,M} = U_e(x_i)$ by applying the top boundary condition. In this way, a complete profile is calculated at $x = x_i + \Delta x$. Observe how easily the boundary and initial conditions entered. The wall boundary condition is used directly in the calculation of all points that are one grid spacing up from the wall (i.e., $m = 2$). The outer boundary condition is used directly in the calculation of all points that are one grid spacing down from the top boundary (i.e., $m = M - 1$).

With this, one step in the downstream marching process is completed, since the solution at $x = x_i + \Delta x$ (i.e., $u_{2,m}$) is now known. The rest of the process is conceptually simple. Consider the newly calculated profile at $x = (x_i + \Delta x)$ as known in the same sense as the given initial profile at $x = x_i$ and step downstream to $x = (x_i + \Delta x) + \Delta x$ in exactly the same fashion as from x_i to $(x_i + \Delta x)$. If you can make two steps, you can certainly make another, and another, and so on, until $x = x_f = (N - 1) \Delta x + x_i$ is reached. One can see that many, many simple, repetitive numerical operations are involved. It is possible to apply this procedure by hand, but it surely would involve more time than any rational person would want to spend. The method is, however, perfectly suited to the digital computer.

Having seen how to achieve a numerical approximation to a parabolic problem, it is natural now to inquire how good the resulting solution might be. It happens that if one blindly applied this procedure to a variety of problems, seemingly good solutions in some cases and obvious nonsense in others would be obtained. One cannot escape mathematics altogether, and the difficulty can be traced to two important items: *round-off error*, and *stability*.

The round-off errors come from the finite number of digits that can be handled in any computer. Clearly, each number must be rounded off at the last digit that can be retained. That statement may sound like a truism, and indeed it is, but the implications are neither small nor altruistic. The problem arises because many, many separate calculations are to be made for any given problem. If the errors accumulate a little at each step, it is possible to reach the point where the errors dominate the solution. Such a situation is termed an *unstable* procedure. If the errors tend to cancel each other, the total error by the end of the problem remains small, and the procedure is called *stable*. *Truncation errors* arise in a different way. Equation (4-45a) can be rearranged and viewed as the first term in a series expansion for $u(y + \Delta y)[u_{n, m+1}]$ from $u(y)[u_{n, m}]$ involving $\partial u/\partial y$. A complete series expansion would also have terms with $(\Delta y)^2(\partial^2 u/\partial y^2)$, and so on. Thus, Eq. (4-45a) represents a *truncated* series. For small Δy it can be an accurate approximation, but the neglected terms will contribute small errors that can accumulate or tend to cancel in the same way as the *round-off* errors.

The stability question can be analyzed mathematically following the procedure developed by von Neumann (1952) for the *heat equation*, Eq. (4-10). The procedure relies on the fact that this equation is linear and uses Fourier analysis. The solution of some physical problems described by the heat equation, and thus also Eq. (4-11), can be written in the form

$$u(x, y) = \xi(x)e^{i\beta y} \qquad (4\text{-}50)$$

so that convenient form can be adopted for the analysis. Here β is a wave number of a periodic function. Since the equation is *linear*, the behavior of an initial *error*, represented by Eq. (4-50), superimposed on a background solution can be analyzed either with or without the background solution present. Clearly, it is more convenient to do so without the background solution. Also, we seek a general result, not the behavior of an error with respect to the solution of a particular problem. Further, since the equation is linear, we can proceed with any single value of β and be assured that the corresponding result for any other β or combination of β's can be found by superposition.

In finite difference form, Eq. (4-50) becomes

$$u_{n, m} = \xi_n e^{i\beta(m-1)\Delta y}$$
$$u_{n, m+1} = \xi_n e^{i\beta m(\Delta y)}$$

$$(4\text{-}51)$$

$$u_{n,\,m-1} = \xi_n e^{i\beta(m-2)\Delta y}$$

$$u_{n+1,\,m} = \xi_{n+1} e^{i\beta(m-1)\Delta y}$$

where $\xi_n = \xi((n-1)\Delta x + x_i)$. Substituting into Eq. (4-48a), we get

$$\xi_{n+1} e^{i\beta(m-1)\Delta y} = Q(\xi_n e^{i\beta m(\Delta y)} + \xi_n e^{i\beta(m-2)\Delta y}) - (2Q-1)\xi_n e^{i\beta(m-1)\Delta y} \quad (4\text{-}52)$$

Divide through by $e^{i\beta(m-1)\Delta y}$ and then ξ_n to get

$$\frac{\xi_{n+1}}{\xi_n} = Q(e^{i\beta(\Delta y)} + e^{-i\beta(\Delta y)}) - (2Q-1)$$

$$= 2Q\,\cos\,(\beta(\Delta y)) - (2Q-1) \quad (4\text{-}53)$$

The last step involves trigonometric identities. In any event, for stability we want $|\xi_{n+1}| \le |\xi_n|$ (i.e., the errors are not to grow). That condition is satisfied for $Q \le \frac{1}{2}$ no matter what the value of β. This is not an abstract result, as the working of Prob. 4.11 will show. The consequences of this restriction can be seen by rewriting it using the definition of Q as

$$\Delta x \le \frac{1}{2}\frac{U_e(\Delta y)^2}{\nu} \quad (4\text{-}54)$$

It usually happens that Eq. (4-54) forces the use of a much smaller Δx than would otherwise seem required to adequately describe the variations of the flow in the x direction. This results in more calculation steps and therefore greater computer cost with no attendant improvement in the solution. Note that one should be cautious about choosing a smaller than necessary value of Δy, since that immediately implies an even smaller value of Δx from Eq. (4-54).

It is possible for a stable procedure that *converges* to a solution as the step sizes are decreased to converge to the solution of a different differential equation than the original. That is, the truncation error may not decrease to zero as Δx and Δy do. By using a Taylor's series, an expression for the truncation error can be developed. For the present explicit method, it is found that the truncation error $\varepsilon_{n,\,m}$ is

$$\varepsilon_{n,\,m} = U_e \frac{\Delta x}{2}\left(\frac{\partial^2 u}{\partial x^2}\right) - \nu\,\frac{(\Delta y)^2}{12}\frac{\partial^4 u}{\partial y^4} + O((\Delta x)^2) + O((\Delta y)^4) \quad (4\text{-}55)$$

Thus. $\varepsilon_{n,\,m} \to 0$ as $\Delta x,\ \Delta y \to 0$, as desired.

A way out of the restriction in Eq. (4-54) can be found with some increase in conceptual complexity. We begin by reconsidering the finite difference approximation to the second derivative with respect to y in Eq. (4-46). The approximation for the second derivative should apply for the region between x and $x + \Delta x$. Thus, one can just as well use

$$\left.\frac{\partial^2 u}{\partial y^2}\right|_{n+1,\,m} \approx \frac{u_{n+1,\,m+1} - 2u_{n+1,\,m} + u_{n+1,\,m-1}}{(\Delta y)^2} \quad (4\text{-}56)$$

In this case, the finite difference representation is formulated using as-yet unknown information at the downstream station that we are stepping toward rather than known information at the station where we are. In the limit as $\Delta x \to 0$, the distinction would be lost, but we are not letting $\Delta x \to 0$, and the distinction is real. Substituting Eqs. (4-56) and (4-47) into (4-11) results in

$$u_{n+1,\, m} = \frac{u_{n,\, m} + Q(u_{n+1,\, m+1} + u_{n+1,\, m-1})}{1 + 2Q} \tag{4-57}$$

This is an expression for the new velocity at one step Δx downstream along the m grid line in terms of the known value before the step on that same grid line and the new values after a Δx step on the $m + 1$ and $m - 1$ grid lines. The information flows as shown in Fig. 4-11(B). Thus, this is an *implicit* relation for one unknown in terms of some of the other unknowns. This is, therefore, called an *implicit finite difference* formulation. Fortunately, we have one equation like Eq. (4-57) for each unknown, so the system is closed. The solution of this system of linear, algebraic equations for all the unknowns at each Δx step downstream involves inverting or decomposing a matrix. Since there may be a large number (approximately 10 to 100) of grid points across the layer, the matrix can be quite large, and that deterred the use of this type of formulation even in the early days of the availability of digital computers. However, the matrix is tridiagonal, that is, all the entries off the main diagonal and one diagonal line to each side are zero. This is because the equation for each unknown involves only two other unknowns, the ones on each side of it, and none of the others. The necessary manipulations are, therefore, not difficult or expensive with modern machines and methods.

With the added complexity of the *implicit* formulation, the reader may well question the utility of the approach. The key lies in the fact that a stability analysis of this formulation proceeding in exactly the same way as for the explicit formulation reveals that the implicit method is unconditionally stable (i.e., it is stable for any value of Q). The only restrictions on the choices of Δx and Δy come from considerations of profile shapes, boundary conditions, and so on, as discussed earlier. Again, this comes down to a matter of cost, and the implicit method is simply cheaper to run for a given problem than the explicit method.

The truncation error for this method is found to be

$$\varepsilon_{n,\, m} = -U_e \frac{(\Delta x)}{2} \left(\frac{\partial^2 u}{\partial x^2} \right) - v \frac{(\Delta y)^2}{12} \left(\frac{\partial^4 u}{\partial y^4} \right) + O((\Delta x)^2) + O((\Delta y)^4) \tag{4-58}$$

which again shows that $\varepsilon_{n,\, m} \to 0$ as $\Delta x, \Delta y \to 0$.

Since the finite difference representation for $\partial^2 u / \partial y^2$ must hold from n to $n + 1$, a logical extension of the simple explicit method ($\partial^2 u / \partial y^2$ evaluated at n) and the simple implicit method above ($\partial^2 u / \partial y^2$ evaluated at $n + 1$) is to take the average of the two, giving $\partial^2 u / \partial y^2$ evaluated at $n + \frac{1}{2}$. Further motivation for that approach is found upon comparing Eqs. (4-55) and (4-58) for the

truncation errors. Clearly, taking the average of the two will cause the cancel-
lation of the term in $\varepsilon_{n,m}$ involving Δx. This method is called the *Crank–
Nicolson method*. It is an implicit method, since it involves unknowns at $n + 1$,
and a stability analysis shows it to be unconditionally stable. Since it is stable
and has improved accuracy (smaller truncation errors) for a given Δx, it is very
popular.

4-6-2 An Explicit Method for the Boundary Layer Equations

From the discussion in Sec. 4-6-1, it should not be surprising to find that
the earliest finite difference codes for the boundary layer equations were based
on an explicit formulation. The explicit method of Wu (1961) has been chosen
for illustrative purposes. The method is presented here in a planar, incompressible form for simplicity and clarity.

The differential equations to be treated are the continuity equation, Eq.
(3-6a), and the momentum equation, Eq. (3-27a), without the unsteady term in
the latter. The parabolic momentum equation is solved with an explicit,
downstream-marching procedure, and the continuity equation is viewed as an
auxiliary equation for determining $v(x, y)$.

The finite difference form of the momentum equation is not difficult to
derive as

$$u_{n,m}\left(\frac{u_{n+1,m} - u_{n,m}}{\Delta x}\right) + v_{n,m}\left(\frac{u_{n,m+1} - u_{n,m-1}}{2(\Delta y)}\right)$$
$$= -\frac{1}{\rho}\left(\frac{p_{n+1} - p_n}{\Delta x}\right) + v\left(\frac{u_{n,m+1} - 2u_{n,m} + u_{n,m-1}}{(\Delta y)^2}\right) \quad (4\text{-}59)$$

Note that a central difference has been used for the first derivative $\partial u/\partial y$ since
it is not in the marching direction. Everything else follows directly from the
material in the preceding section. This equation provides an explicit relation
for calculating the streamwise velocity after a downstream step $u_{n+1,m}$ in terms
of known information at the station before the step.

$$u_{n+1,m} = \frac{v(\Delta x)}{u_{n,m}(\Delta y)^2}(u_{n,m+1} + u_{n,m-1}) - \left(2\left(\frac{v(\Delta x)}{u_{n,m}(\Delta y)^2}\right) - 1\right)u_{n,m}$$
$$- \frac{1}{\rho u_{n,m}}(p_{n+1} - p_n) - \frac{v_{n,m}}{u_{n,m}}\left(\frac{\Delta x}{\Delta y}\right)\left(\frac{u_{n,m+1} - u_{n,m-1}}{2}\right) \quad (4\text{-}59a)$$

Observe the similarity with Eq. (4-48a) for the *model* equation. In this case
there are two additional terms from the pressure gradient and the second term
in the convective derivative involving v. Also, note that the factor involving the
step sizes [Q in Eq. (4-48a)] is no longer a constant across the layer at any x
station, since it now contains $u_{n,m}$ rather than U_e.

The finite difference form of the simpler, differential continuity equation proves a little more difficult to develop as a result of the added complication of coupling between two equations which we have not considered before. A straightforward formulation would be

$$\frac{u_{n+1,\,m} - u_{n,\,m}}{\Delta x} + \frac{v_{n,\,m+1} - v_{n,\,m-1}}{2(\Delta y)} = 0 \qquad (4\text{-}60)$$

This equation is to be used to calculate the transverse velocity after a down-stream step $v_{n+1,\,m}$. Looking at Eq. (4-60), one can see that none of those quantities appear, so this formulation will not serve. If the $\partial v/\partial y$ term is approximated in terms of quantities after the step

$$\frac{\partial v}{\partial y} \approx \frac{v_{n+1,\,m+1} - v_{n+1,\,m-1}}{2(\Delta y)} \qquad (4\text{-}61)$$

an equation for v at every other grid point on the new profile results. It would then be necessary to interpolate for the values at the missing points. To obtain good accuracy, this requires the use of a very small step size, Δy, since $\partial v/\partial y$ is large near the wall. The final scheme developed and actually employed involved a backward difference for $\partial v/\partial y$.

$$\left.\frac{\partial v}{\partial y}\right|_{n+1,\,m-1/2} \approx \frac{v_{n+1,\,m} - v_{n+1,\,m-1}}{\Delta y} \qquad (4\text{-}62)$$

This however, can be interpreted as a central difference about a location halfway between two grid points $(n+1,\,m)$ and $(n+1,\,m-1)$, that is, at the point $(n+1,\,m-\frac{1}{2})$ shown as a cross on Fig. 4-12. The expression used for $\partial u/\partial x$ in Eq. (4-60) can be viewed as a central difference around $(n+\frac{1}{2},\,m)$, shown as a triangle on Fig. 4-12. That is rather far from $(n+1,\,m-\frac{1}{2})$, so for consistency, it is now necessary to write a corresponding form for $\partial u/\partial x$.

$$\left.\frac{\partial u}{\partial x}\right|_{n+1,\,m-1/2} = \frac{1}{2}\left[\left.\frac{\partial u}{\partial x}\right|_{n+1,\,m} + \left.\frac{\partial u}{\partial x}\right|_{n+1,\,m-1}\right]$$

$$\approx \frac{1}{2}\left[\frac{u_{n+1,\,m} - u_{n,\,m}}{\Delta x} + \frac{u_{n+1,\,m-1} - u_{n,\,m-1}}{\Delta x}\right] \qquad (4\text{-}63)$$

The next matter of importance is stability, since this is an explicit method. The system of equations is nonlinear, so the method of von Neumann cannot be rigorously applied. The *circuit analogy* method of Karplus (1958) can be applied on an ad hoc basis to this nonlinear system. The analogy comes from the fact that the voltage or current distribution of a network of electrical resistors arranged in a regular pattern can be described by an equation of the form

$$a(\phi_{n,\,m+1} - \phi_{n,\,m}) + b(\phi_{n,\,m-1} - \phi_{n,\,m})$$

$$+ c(\phi_{n+1,\,m} - \phi_{n,\,m}) + d(\phi_{n-1,\,m} - \phi_{n,\,m}) = 0 \qquad (4\text{-}64)$$

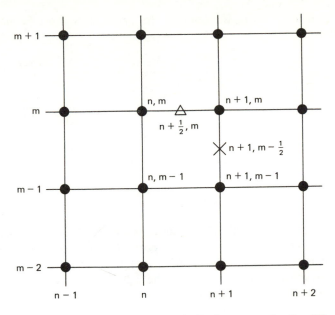

Figure 4-12 Schematic of the locations of effective centers for the differencing scheme of Wu.

where a is positive and ϕ is any dependent variable. It can be proven that if (1) all the coefficients are positive or (2) some are negative, and the sum is negative, then the system is stable. Since all the finite difference approximations discussed so far can be put in the form of Eq. (4-64), the analogy can be used to infer stability of the finite difference formulation.

For the formulation under discussion, the equations can be rearranged to give

$$a = -\frac{\Delta x}{2(\Delta y)}\, v_{n,\,m} + \frac{v(\Delta x)}{(\Delta y)^2}$$

$$b = \frac{\Delta x}{2(\Delta y)}\, v_{n,\,m} + \frac{v(\Delta x)}{(\Delta y)^2} \qquad (4\text{-}65)$$

$$c = -\, u_{n,\,m}$$

$$d = 0$$

where

$$a > 0 \qquad \text{if } v_{n,\,m} < 0$$

or

$$\Delta y < \frac{2v}{v_{n,\,m}} \qquad \text{if } v_{n,\,m} > 0 \qquad (4\text{-}66)$$

Since $u_{n,m} > 0$, $c < 0$, and the sum of the coefficients must be negative for stability; that is,

$$\frac{2\nu(\Delta x)}{(\Delta y)^2} - u_{n,m} < 0 \tag{4-67}$$

or

$$\frac{\nu(\Delta x)}{u_{n,m}(\Delta y)^2} < \frac{1}{2} \tag{4-67a}$$

leading to

$$\Delta x < \frac{1}{2}\frac{u_{n,m}(\Delta y)^2}{\nu} \tag{4-68}$$

This result is nearly identical to that rigorously derived for the *model* equation in Sec. 4-6-1, $\nu(\Delta x)/(U_e(\Delta y)^2) < \frac{1}{2}$. The important difference is that the u in the denominator is a variable in Eq. (4-67a) that is always less than U_e, so this requirement is more restrictive. Indeed, the value of $u_{n,m}$ at the first grid point out from the wall will generally be a small fraction of U_e, and that sets the restriction on Δx for the whole layer.

Earlier, we found the condition in Eq. (4-66) which must be met to keep $a > 0$. That must still be obeyed.

This method is simple and quite successful. Even with the step-size restrictions described above, accurate calculations can be made for planar, incompressible, constant-property flows for a matter of pennies in computer cost.

4-6-3 An Implicit Method
for the Boundary Layer Equations

An early implicit procedure for the boundary layer equations was developed by Parr (1963). Parr used

$$\frac{\partial^2 u}{\partial y^2} \approx \frac{1}{2}\left[\frac{u_{n+1,m+1} - 2u_{n+1,m} + u_{n+1,m-1}}{(\Delta y)^2} + \frac{u_{n,m+1} - 2u_{n,m} + u_{n,m-1}}{(\Delta y)^2}\right] \tag{4-69}$$

which is the Crank–Nicolson method. It can be noted that the first term in the brackets is an *implicit* form and the second is an *explicit* form. Parr found that an unconditionally stable procedure also resulted for the boundary layer equations. This *implicit* method is successful, and it is even cheaper to run than the corresponding explicit method discussed earlier.

4-6-4 Transformations and Other Matters

With the material in Secs. 4-6-2 and 4-6-3, the reader has already seen the primary methods used for boundary layer problems. Further extensions of

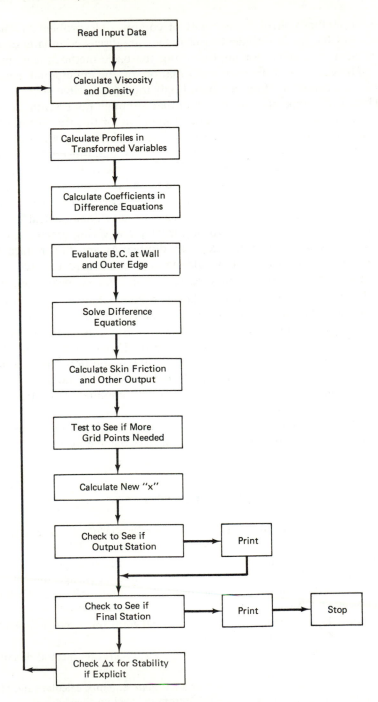

Figure 4-13   Flow chart for a typical boundary layer computer code.

these basic methods involve treatment of compressible problems (which are discussed in Chapter 5), turbulent cases (which are discussed in later chapters), more efficient procedures for implementing the basic methods (e.g., manipulating matrices), and/or prior transformation of the differential equations. Some of these transformations are useful only for compressible flows.

There is one type of transformation that although not essential, is helpful for virtually all methods and cases. The motivation for this transformation can be seen in Fig. 4-10. The thickness of the region of interest $\delta(x)$ grows along the surface, so the computational region N points high must either also grow or be set much too large near the upstream boundary. The latter strategy is obviously costly, since it implies solving the boundary layer equations at many points out in the essentially inviscid flow. It is perfectly feasible to have N increase as $\delta(x)$ increases. It is only necessary to have some logical test in the code. The value of N may grow to be rather large, giving greater resolution than is needed. One can then increase (say double) Δy and thus reduce N, but this is an extra complication. It would be very convenient to have a transformed, transverse coordinate that remains constant, or nearly so, as $\delta(x)$ grows. If the flow were over a flat plate, we know that $\delta \sim \sqrt{x}$. For other bodies, that dependence would crudely approximate the growth rate. Thus, we are led to a new, transverse variable $\sim y/\sqrt{x}$. But this is the form of the similarity variable, η, so why not simply use it? Don't be confused; the discussion is *not* being restricted to cases with similar solutions. We are merely making the acceptable mathematical transformation of variables (x, y) to (x, η). This transformation is often combined with compressibility transformations, as shown in Chapter 5.

A typical computer code for solving boundary layer problems by the methods described here might have a *flowchart* like the one shown in Fig. 4-13.

PROBLEMS

4.1. Develop an expression for the axial velocity profile $u(y)$ for fully developed, laminar flow between two parallel plates separated by a distance $2H$.

4.2. Consider the axial laminar flow between two, parallel *porous* plates separated by a distance $2H$. There is a transverse velocity, v_w, into the channel through the top plate and out through the bottom plate at the same value, v_w. Develop an expression for the axial velocity profile, $u(y)$, for fully developed flow.

4.3. Substitute Eqs. (4-13) and (4-14) and the trial form for η into Eq. (4-12) and determine A, B, and C to give Eq. (4-16).

4.4. Write and test a computer program for the Piercy–Preston method described in Sec. 4-3-1.

4.5. For air at 1 atm and 25°C flowing over a flat plate at 5 m/s, calculate and plot the v velocity profile at a distance $x = 0.3$ m from the leading edge.

4.6. Show that $(\delta^*/\tau_w)\, dp/dx$ represents the ratio of pressure force to wall friction force on the fluid in a boundary layer. Show that it is constant for any of the Falkner–Skan *wedge* flows.

4.7. Water flowing through a 3-m section of a long pipe of 1 cm diameter at a volume flow rate of 15 cm^3/s is cooled from 100°C to 30°C. The inside wall temperature of the pipe is 20°C. What is the heat transfer coefficient?

4.8. Consider airflow over a flat plate for two cases: one with T_w = constant and the other with $(T_w - T_e) \sim x^2$. Compare the thickness of the thermal boundary layers for the two cases at the same station, x.

4.9. The inside of a pipe is covered with a material that is highly absorbent of CO_2. The inside diameter is 4 cm. A mixture of 5% CO_2 in air enters the pipe at 0.5 m/s at standard temperature and pressure. The CO_2 is so strongly absorbed that the mass fraction of CO_2 on the wall may be taken as zero. What is the *bulk* concentration of CO_2 after 10 m of pipe assuming fully developed flow over the whole length?

4.10. Write and test a computer program for one axial step, Δx, using the explicit method and the *model* equation as described in Sec. 4-6-1. Take the initial profile, at $x = x_i$, as Eq. (2-45), with U_e = constant = 3 m/s, $\delta(x_i)$ = 0.5 cm, and the fluid as air at 1 atm and 25°C. Use $\Delta y = \delta/20$ and $\Delta x = \delta/2$.

4.11. To test the importance of the stability criteria developed in Sec. 4-6-1, examine the behavior of an initial error represented by $u_{1,1} = 1.0$ and $u_{1,m} = 0$ for all $m \neq 1$. Using Eq. (4-48a), make calculations for $n = 2, 3$ and $m = 1, 2, 3, 4, 5$ with $Q = 0.25, 0.5,$ and 0.6. Take the flow as symmetrical about the $m = 1$ line.

4.12. Redo Prob. 4.10 using the implicit formulation.

4.13. Perform a stability analysis of the implicit finite difference formulation for the *model* equation, Eq. (4-57), using the method used for the explicit formulation in Sec. 4-6-1.

4.14. Perform a stability analysis of the simple, explicit, finite difference formulation for the *model* equation, Eq. (4-48a), using the Karplus method.

4.15. Derive a finite difference formulation of the *model* equation, Eq. (4-11), based on the Crank–Nicholson method.

4.16. Assume that the following equation is a suitable approximate equation for boundary layer problems with suction or injection [i.e., $v_w(x) \neq 0$]:

$$U_e \frac{\partial u}{\partial x} + v_w \frac{\partial u}{\partial y} = v \frac{\partial^2 u}{\partial y^2}$$

Develop an explicit finite difference approximation for the solution of this equation. Perform a stability analysis and state the results.

5

COMPRESSIBLE
LAMINAR BOUNDARY LAYERS

5-1 INTRODUCTION

For a steady, compressible, laminar flow over a planar or axisymmetric body, the equations to be treated are the steady forms of Eqs. (3-27), (3-28), and (3-51) [and (3-61) if there is more than one species]. Looking at this system and limiting the discussion to a single species for the moment, one sees that the influence of compressibility is first contained directly in the density terms in the continuity equation, Eq. (3-28), and more passively as a variable coefficient in the momentum (3-27) and energy (3-51) equations. The second influence of compressibility is to produce temperature variations that are too large to permit the assumption of constant properties, μ and k. Actually, the main concern here is with the viscosity, since we may use the definition of the Prandtl number to write

$$k = \frac{\mu c_p}{\mathrm{Pr}} \tag{5-1}$$

The advantage of this will become clearer if the reader recalls that the Prandtl number is nearly constant for most gases over a wide range of temperature (see Fig. 1-3 and the tables in Appendix A). Further, it is common to use the energy equation written in terms of the enthalpy, Eq. (3-50), in compressible flow problems. Then one writes

$$-\frac{\partial q_y}{\partial y} = \frac{\partial}{\partial y}\left(k\,\frac{\partial T}{\partial y}\right) = \frac{\partial}{\partial y}\left(\frac{k}{c_p}\,\frac{\partial h}{\partial y}\right) = \frac{\partial}{\partial y}\left(\frac{\mu}{\mathrm{Pr}}\,\frac{\partial h}{\partial y}\right) \tag{5-2}$$

In the simplest cases, therefore, one may say that the added complexity with compressible, laminar boundary layer problems is centered on variable ρ and μ and various, but essentially constant, values of Pr. We know that $\mu = \mu(T)$, and an equation of state will give, in general, $\rho = \rho(T, p)$. However, the pressure is constant across the layer, so we have really to contend only with density variations produced by temperature variations in the boundary layer for single-species gas flows.

5-2 THE ADIABATIC WALL TEMPERATURE

A section with this heading may seem surprising, since, by Eq. (2-55), the wall temperature for adiabatic (no heat transfer) conditions would appear to be just $T_w = T_e$. That equation is, however, suitable only to low-speed cases. At high speeds, where compressibility becomes important, a new phenomenon occurs. By definition, a high-speed flow has appreciable kinetic energy, and that kinetic energy can be dissipated into heat by friction within the boundary layer. This process is modeled by the term $\mu(\partial u/\partial y)^2$ in the energy equation. The kinetic energy in the flow is simply the difference between the total (stagnation) and static temperatures

$$T_{\text{kinetic}} = T_t - T = \frac{V^2}{2c_p} \qquad (5\text{-}3)$$

If the wall is adiabatic, the temperature that the wall attains at equilibrium will depend on how much of this kinetic energy is *recovered* on the wall. This is expressed as a *recovery factor*, r, defined as

$$T_{\text{aw}} = T_e + r\,\frac{U_e^2}{2c_p} \qquad (5\text{-}4)$$

The value of r is generally less than, but near, unity for gases. More will be said about that later. In any event, a film coefficient for high-speed flow must be based on

$$q_w = h(T_w - T_{aw}) \qquad (5\text{-}5)$$

5-3 THE REFERENCE TEMPERATURE METHOD

In view of the discussion in Sec. 5-1, it is reasonable to ask if there might be some value of the temperature, say between T_e and T_w, where the density and physical properties could be evaluated and used in the available constant-density, constant-property solutions to provide an adequate approximation to the actual, variable-density, variable-property flow. The answer, at least for gases, is *yes*, and that value of the temperature is called the *reference temper-*

ature, T*. Here we quote the relation due to Eckert (1956):

$$T^* = T_e + 0.5(T_w - T_e) + 0.22(T_{aw} - T_e) \tag{5-6}$$

An equivalent relation with the same high degree of accuracy of results has not been developed for liquids. Also, for lower-speed flows that may still involve large temperature differences, the last term in Eq. (5-6) disappears.

With this concept, all the solutions in Chapters 2 and 4 can be carried over to compressible flow problems. For example, we may rewrite Eq. (2-59), with $x_0 = 0$, as

$$St^* = 0.332(Pr^*)^{-2/3}(Re^*)^{-1/2} \tag{5-7}$$

where the * superscript denotes that all physical properties and the density are to be evaluated at $T = T^*$. This formula can then be used to give the film coefficient h that can be used with Eq. (5-5) for high-speed flow over a flat plate.

5-4 THE SPECIAL CASE OF PRANDTL NUMBER UNITY

We saw in Sec. 4-4-2 that the case $Pr = 1$ was a special one giving identical nondimensional profiles for velocity and temperature in low-speed flow. This matter can be pursued further for high-speed flows. Although this is a special situation not generally found in the real world, many gases have $Pr \approx 0.7$ (see Appendix A), which is not far from unity, and the results of studying the $Pr = 1$ case are quite informative.

For our purposes here, take the energy equation as in Eq. (3-48) with the heat transfer term written as in Eq. (5-2) and the shear as $\mu \, \partial u/\partial y$. Also, restrict the discussion to planar, steady flow to arrive at

$$\left(\rho u \frac{\partial}{\partial x} + \rho v \frac{\partial}{\partial y}\right)\left(h + \frac{u^2}{2}\right) = \frac{\partial}{\partial y}\left[\mu\left(\frac{\partial}{\partial y}\left(\frac{h}{Pr} + \frac{u^2}{2}\right)\right)\right] \tag{5-8}$$

Clearly, if $Pr = 1$, a solution to this equation is

$$h + \frac{u^2}{2} = \text{constant} \tag{5-9}$$

The *constant* can be evaluated from a boundary condition, say at the wall where $u \equiv 0$. A suitable, simple case will be an insulated wall with no heat flow, and the result is

$$h + \frac{u^2}{2} = h_{aw} = \text{constant} = h_e + \frac{U_e^2}{2} \tag{5-10}$$

or

$$T + \frac{u^2}{2c_p} = T_{aw} = \text{constant} = T_e + \frac{U_e^2}{2c_p} \tag{5-10a}$$

This result is known as the Busemann (1931) energy integral.

This analysis was generalized by Crocco (1932). The general assumption $h = h(u)$ is substituted into the steady form of Eq. (3-50) with the heat transfer term as in Eq. (5-2) and the shear as usual to give

$$\rho \frac{dh}{du} \left(u \frac{\partial u}{\partial x} + v \frac{\partial u}{\partial y} \right) = \frac{\partial}{\partial y} \left(\frac{\mu}{\text{Pr}} \frac{dh}{du} \frac{\partial u}{\partial y} \right) + u \frac{dp}{dx} + \mu \left(\frac{\partial u}{\partial y} \right)^2 \qquad (5\text{-}11)$$

where we have used

$$\frac{\partial h}{\partial x} = \frac{dh}{du} \frac{\partial u}{\partial x}$$

$$\frac{\partial h}{\partial y} = \frac{dh}{du} \frac{\partial u}{\partial y} \qquad (5\text{-}12)$$

The first term on the right-hand side of Eq. (5-11) can be expanded with the product rule to yield

$$\frac{dh}{du} \frac{\partial}{\partial y} \left(\frac{\mu}{\text{Pr}} \frac{\partial u}{\partial y} \right) + \frac{\mu}{\text{Pr}} \frac{d^2h}{du^2} \left(\frac{\partial u}{\partial y} \right)^2 \qquad (5\text{-}13)$$

The last term in Eq. (5-13) comes from continuing the logic shown in Eq. (5-12). Now take Eq. (5-11) with Eq. (5-13) and combine it with the steady momentum equation, Eq. (3-25), multiplied by dh/du, which results in

$$-\frac{dp}{dx} \left[\frac{dh}{du} + u \right] + \frac{dh}{du} \left[\frac{\partial}{\partial y} \left(\mu \frac{\partial u}{\partial y} \right) - \frac{\partial}{\partial y} \left(\frac{\mu}{\text{Pr}} \frac{\partial u}{\partial y} \right) \right]$$
$$= \frac{\mu}{\text{Pr}} \left(\frac{\partial u}{\partial y} \right)^2 \left[\frac{d^2h}{du^2} + \text{Pr} \right] \qquad (5\text{-}14)$$

There are two cases which admit to simple solutions by inspection. Both must have $\text{Pr} = 1$.

The first case arises for

$$\frac{dh}{du} + u = 0 \qquad (5\text{-}15)$$

With this, Eq. (5-14) becomes

$$\frac{d^2h}{du^2} + 1 = 0 \qquad (5\text{-}16)$$

leading to

$$h = -\frac{u^2}{2} + \text{constant} \qquad (5\text{-}9)$$

which is the earlier result. By differentiating and evaluating at the wall, $u = 0$:

$$\left. \frac{\partial h}{\partial y} \right|_w = -\overset{o}{u}_w \left. \frac{\partial u}{\partial y} \right|_w = 0 \qquad (5\text{-}17)$$

it can be seen that this requires no heat transfer if it is noted that

$$\frac{\partial h}{\partial y}\Big|_w = c_p \frac{\partial T}{\partial y}\Big|_w = -\frac{c_p}{k} q_w \tag{5-18}$$

It is important to observe here, however, that this solution holds for arbitrary dp/dx.

The second case emerges under the restriction of $dp/dx \equiv 0$, which again leads to Eq. (5-16) but without Eq. (5-15). Simple integration yields in this case

$$h = -\frac{u^2}{2} + C_1 u + C_2 \tag{5-19}$$

The constants, C_1 and C_2, can be found from the conditions

$$h(U_e) = h_e$$
$$h(0) = h_w \tag{5-20}$$

The final result is

$$h = -\frac{u^2}{2} + \left(h_e + \frac{U_e^2}{2} - h_w\right)\frac{u}{U_e} + h_w \tag{5-21}$$

or

$$T = -\frac{u^2}{2c_p} + \left(T_e + \frac{U_e^2}{2c_p} - T_w\right)\frac{u}{U_e} + T_w \tag{5-21a}$$

Differentiating the first expression and evaluating at the wall gives

$$\frac{\partial h}{\partial y}\Big|_w = -u_w^0 \frac{\partial u}{\partial y}\Big|_w + \left(\frac{h_e + U_e^2/2 - h_w}{U_e}\right)\frac{\partial u}{\partial y}\Big|_w \tag{5-22}$$

which is not zero. This case thus allows wall heat transfer under the restriction of a zero pressure gradient.

The relation in Eq. (5-21a) can be used to show an important point about very high speed boundary layer flows. Consider the case of flow over a constant-pressure region of a body moving at $M_e = 8$ in the atmosphere at sea level, $T_e = 300$ K. Using isentropic relations, we find that

$$\left(\frac{T}{T_t}\right)_e = 0.0725 \tag{5-23}$$

Thus, $T_{t,\,e} \equiv T_e + U_e^2/2c_p = 4140$ K, if it is assumed for the purposes of this illustrative example only that air behaves as a perfect gas under these conditions. The maximum tolerable surface temperature for most practical applications is about 1000 K. We now seek the maximum value of the static temperature in the boundary layer. The working of Prob. 5.3 will show that the maximum static temperature occurs where

$$\left.\frac{u}{U_e}\right|_{T=T_{max}} = \frac{T_{t,e} - T_w}{U_e^2/c_p} \tag{5-24}$$

For the hypothetical case under discussion, we find that $T_{max} = 1640$ K, which is much higher than either T_e or T_w. This shows the existence of a static *temperature peak* in the boundary layer for high-Mach-number flows. This is very important for reentry vehicles, since it leads to dissociation and ionization of the air in the boundary layer, producing the communications *blackout* familiar to all avid TV watchers of the space program.

Finally, we note from Eq. (5-10a) that the adiabatic wall temperature T_{aw} for Pr = 1 equals $T_{t,e}$. That says that for Pr = 1, the recovery factor r is also unity.

5-5 THE RECOVERY FACTOR FOR NONUNITY PRANDTL NUMBER

A useful, rather general result can be found by analyzing a simple, seemingly restrictive problem—the *plate thermometer*. This is the case of an adiabatic, flat plate suspended in a high-speed fluid stream. The temperature measured by a thermometer (e.g., a thermocouple) embedded in the plate will be, by definition, the adiabatic wall temperature T_{aw}. Knowing this, Pr, and $T_{t,e}$, it will be possible to find $r = r(Pr)$.

Pohlhausen (1921a) treated this problem under the constant-property assumption, but including, of course, the frictional heating term, $\mu(\partial u/\partial y)^2$, in the energy equation. We use the steady form of Eq. (3-52) with $dp/dx = 0$, written as

$$u\frac{\partial T}{\partial x} + v\frac{\partial T}{\partial y} = \frac{v}{Pr}\frac{\partial^2 T}{\partial y^2} + \frac{v}{c_p}\left(\frac{\partial u}{\partial y}\right)^2 \tag{5-25}$$

The boundary conditions are

$$y = 0, \, x \geq 0: \quad \frac{\partial T}{\partial y} = 0$$

$$y \rightarrow \infty, \text{ all } x: \quad T(x, y) \rightarrow T_e$$

It is helpful to introduce a dimensionless *excess temperature*

$$\Theta_r \equiv \frac{T - T_e}{U_e^2/(2c_p)} \tag{5-26}$$

The similarity variable η and the representation of the velocity field in terms of $f(\eta)$ from Secs. 4-3-1 and 4-4-2 can also be used here. With all of that and Eq.

(5-26), Eq. (5-25) becomes

$$\Theta_r'' + \Pr \cdot f \Theta_r' + \frac{P_1}{2}(f'')^2 = 0 \tag{5-27}$$

The boundary conditions are

$$\Theta_r'(0) = 0$$
$$\lim_{\eta \to \infty} \Theta_r(\eta) = 0 \tag{5-28}$$

The solution of Eq. (5-27) subject to Eq. (5-28) can be found directly by the method of *variation of parameters*. Our primary interest here is in the value of the temperature at the wall, from which the recovery factor may be found. The result is

$$r = \frac{\Pr}{2} \int_0^\infty \exp\left(-\Pr \int_0^\eta f\, d\eta\right)\left[\int_0^\eta (f'')^2 \exp\left(\Pr \int_0^\eta f\, d\eta\right) d\eta\right] d\eta$$

$$\approx \sqrt{\Pr} \qquad \text{for } 0.5 \le \Pr \le 5.0 \tag{5-29}$$

The key result $r = \sqrt{\Pr}$ for laminar flow has been found to hold under

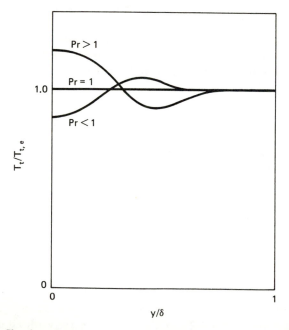

Figure 5-1 Sketches of total temperature profiles in a high-speed boundary layer for Prandtl number greater than, equal to, and less than unity.

much more general conditions than those for this simple flow problem. Indeed, it is reputed to have been a highly classified secret on both sides during World War II.

Sketches of typical total temperature profiles across a boundary layer on an adiabatic wall for $Pr > 1$, $Pr = 1$, and $Pr < 1$ are shown in Fig. 5-1. In Sec. 5-4 it was seen that $T_t = \text{constant} = T_{t, e}$ for $Pr = 1$ [see Eq. (5-10a)]. For $Pr < 1$, the recovery factor r is also less than unity [see Eq. (5-29)]. This means that $T_{aw} = T(x, 0) = T_t(x, 0) < T_{t, e}$. Since the flow is adiabatic, the total thermal energy in the boundary layer must remain constant. Thus, if one region of the flow has $T_t < T_{t, e}$, some other region must have $T_t > T_{t, e}$, as shown in Fig. 5-1. This picture is reversed for $Pr > 1$, where $T_{aw} > T_{t, e}$. All of this is a result of the fact that the Prandtl number is the ratio of diffusion coefficients for momentum and thermal energy.

5-6 COMPRESSIBILITY TRANSFORMATIONS

With the increased complexity of the equations of motion for compressible (variable-density), variable-property flows, it was natural to seek ways of rigorously extending the material in hand for constant-density, constant-property flows to those cases. Ways were sought to transform a compressible boundary layer problem into an equivalent incompressible problem. The existing methods could be applied to the incompressible problem, and then the solution could be transformed back to give a solution for the original compressible problem. This quest ended in success, with some restrictions, and we will discuss the first and then the latest of the various transformations.

The first *compressibility transformation* seems to have been developed independently by Howarth (1948) and Dorodnitsyn (1942). The basic idea is to introduce a distortion of the transverse coordinate y using the density as a *weighting factor*

$$Y \equiv \int_0^y \frac{\rho}{\rho_e} \, dy \qquad (5\text{-}30)$$

The streamwise coordinate x is left undistorted. The velocity components are now written in terms of a compressible stream function [see Eq. (3-10)] by

$$u = \frac{\partial \psi}{\partial Y}; \qquad v = -\frac{\rho_e}{\rho} \left[\frac{\partial \psi}{\partial x} + u \left(\frac{\partial Y}{\partial x} \right)_y \right] \qquad (5\text{-}31)$$

Substitution into the steady, momentum equation for flow over a flat plate, $dp/dx = 0$, results in

$$\frac{\partial \psi}{\partial Y} \frac{\partial^2 \psi}{\partial x \, \partial Y} - \frac{\partial \psi}{\partial x} \frac{\partial^2 \psi}{\partial Y^2} = \frac{\partial}{\partial Y} \left(\frac{\rho \mu}{\rho_e^2} \frac{\partial^2 \psi}{\partial Y^2} \right) \qquad (5\text{-}32)$$

This equation in the (x, Y) plane is very similar to that for constant-density, constant-property flow in the (x, y) plane [Eq. (3-29) without the unsteady and pressure gradient terms]. The equivalence becomes complete *if* one takes $\rho\mu = $ constant. Since for a given pressure $\rho \sim T^{-1}$, this requires $\mu \sim T$, which is a stronger variation with temperature than is actually observed for air and other common gases. Nonetheless, this approach does provide a method for treating compressible boundary layer flows. One can, for example, take the Blasius solution (Sec. 4-3-1) to be applicable in the (x, Y) plane. The inversion of the transformation requires evaluating

$$y = \int_0^Y \frac{\rho_e}{\rho}\, dY \qquad (5\text{-}33)$$

which requires knowledge of $\rho(x, Y)$. In general, this must come from a solution of the energy equation in the (x, Y) plane to obtain $T(x, Y)$. It is not necessary, but one can also make assumptions that permit the use of a Crocco integral, Eq. (5-10a) for an adiabatic wall or (5-21a), both for $\mathrm{Pr} = 1$. This gives $T(x, Y)$ directly in terms of the known $u(x, Y)$. The results for adiabatic wall cases are shown in Fig. 5-2. Note that the attendant high temperature and thus low density near the wall (small y) stretch out the inner, nearly linear portion of the Blasius profile at high Mach numbers.

At this point, we jump over a great deal of work by several researchers to come to the most general transformation, which is usually termed the Levy–Lees (1956) transformation, $(x, y) \rightarrow (s, \bar{\eta})$.

$$s \equiv \int_0^x \rho_e U_e \mu_e r_0^{2j}\, dx$$

$$\bar{\eta} \equiv \frac{\rho_e U_e r_0^j}{\sqrt{2s}} \int_0^y \frac{\rho}{\rho_e}\, dy \qquad (5\text{-}34)$$

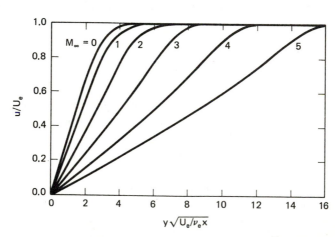

Figure 5-2 Velocity profiles on an insulated flat plate for laminar flow with Prandtl number unity and viscosity proportional to temperature. (From Crocco, 1946.)

where r_0 is the local body radius (see Fig. 3-2). This transformation clearly contains elements of the Howarth–Dorodnitsyn transformation, the Blasius similarity variable, and the effects of an axisymmetric body ($j = 1$) or a planar flow ($j = 0$). Thus, it will be useful for compressible, similar flows. It is also useful for compressible, nonsimilar flows to be treated by numerical methods since it keeps the growth of the computational region in the transverse direction under control (see Sec. 4-6-4). Indeed, in recent times, the latter is by far the greatest advantage of the Levy–Lees transformation.

The results of applying this transformation to the two main operators found in the boundary layer equations are

$$\rho u \frac{\partial(\cdot)}{\partial x} + \rho v \frac{\partial(\cdot)}{\partial y} = \rho U_e^2 \rho_e \mu_e r_0^{2j} \left(f' \frac{\partial(\cdot)}{\partial s} - f' \frac{\partial \bar{\eta}}{\partial s} \frac{\partial(\cdot)}{\partial \bar{\eta}} - \frac{f}{2s} \frac{\partial(\cdot)}{\partial \bar{\eta}} \right)$$

$$\frac{\partial}{\partial y} \left([\cdot] \frac{\partial(\cdot)}{\partial y} \right) = \frac{\rho U_e^2 r_0^{2j}}{2s} \frac{\partial}{\partial \bar{\eta}} \left(\rho[\cdot] \frac{\partial(\cdot)}{\partial \bar{\eta}} \right)$$

(5-35)

where $u(s, \bar{\eta}) = U_e(s) f'(\bar{\eta})$.

5-7 EXACT SOLUTIONS FOR COMPRESSIBLE FLOW OVER A FLAT PLATE

The first solutions for high-speed airflow over a flat plate with a realistic viscosity law and without the restriction of $\text{Pr} = 1$ were obtained by Crocco (1946). He used a transformation of his own devising and reduced the problem to the simultaneous solution of two ordinary differential equations. He made other simplifying assumptions and then solved the problem by the method of successive approximations. The solution was obtained without the aid of a modern computer, and the work can only be described as by Crocco himself as "extremely laborious." With all of this, however, the solutions were essentially exact and their attainment was a landmark event at the time.

Very extensive calculations using the Crocco method were presented by Van Driest (1952). He used the Sutherland viscosity law and $\text{Pr} = 0.75$, and both adiabatic wall and heat transfer (with various T_w/T_e) cases were studied over a wide range of Mach number, $0 \leq M_e \leq 20$. Actually, the results at the higher Mach numbers cannot be viewed as exact, since they were obtained based on a perfect gas assumption. This is untenable for practical cases of high M_e, certainly for $M_e > 6$ since dissociation becomes important. Still Van Driest's results, as a whole, are worthy of careful study, since they cover such wide ranges of the important parameters. In Fig. 5-3, one can see the variation of skin friction coefficient with Mach number. Note the strong effect of T_w/T_e at low Mach number and the important difference between adiabatic and heat transfer cases. Velocity profiles for adiabatic cases are shown in Fig. 5-4. They are generally similar to those in Fig. 5-2, where $\text{Pr} = 1$. Again, there is an extended linear portion of the profile near the wall, and the dimensionless

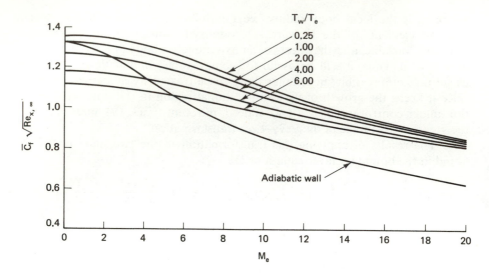

Figure 5-3 Mean skin friction coefficient for laminar flow over a flat plate with Pr = 0.75 and the Sutherland viscosity law. (From Van Driest, 1952.)

boundary layer thickness grows rapidly with edge Mach number. Temperature profiles for the same cases are plotted in Fig. 5-5.

Velocity and temperature profiles for a heat transfer (to the wall) case with $T_w/T_\infty = 1.0$ are given in Figs. 5-6 and 5-7. The velocity profiles have more curvature near the wall than those for the adiabatic cases in Fig. 5-4. The

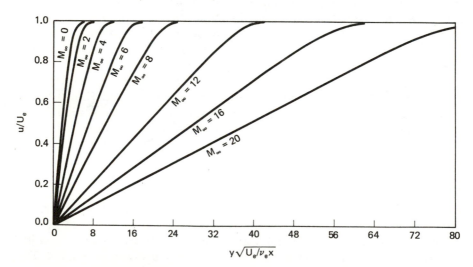

Figure 5-4 Velocity profiles on an insulated flat plate for laminar flow with Pr = 0.75 and the Sutherland viscosity law. (From Van Driest, 1952.)

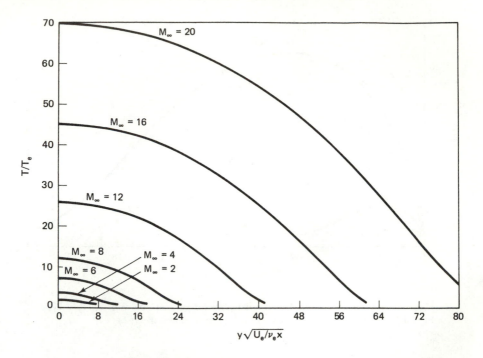

Figure 5-5 Temperature profiles on an insulated flat plate for laminar flow with Pr = 0.75 and the Sutherland viscosity law. (From Van Driest, 1952.)

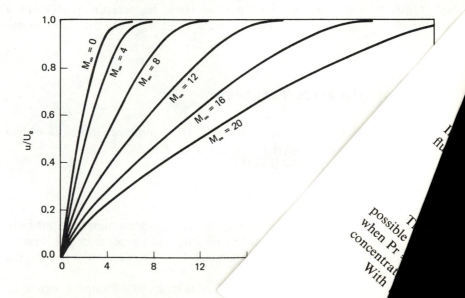

Figure 5-6 Velocity profiles on a flat pl. Pr = 0.75, and the Sutherland viscosity law. (F.

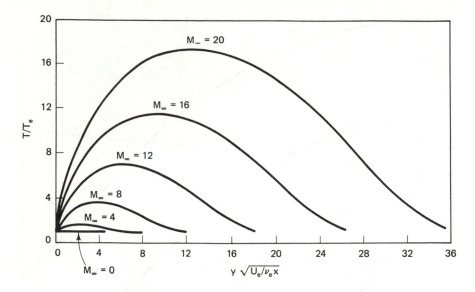

Figure 5-7   Temperature profiles on a flat plate for laminar flow with $T_w/T_e = 1.0$, $Pr = 0.75$, and the Sutherland viscosity law. (From Van Driest, 1952.)

static temperature profiles in Fig. 5-7 show the peaks within the layer discussed earlier. Clearly, very high temperatures can be reached in the layer at high Mach numbers. In Fig. 5-8 we show static temperature profiles as a function of T_w/T_e for $M_e = 4.0$. For hot walls, $T_w \geq T_{aw}$, and the peaks disappear.

5-8 FLOWS WITH MASS TRANSFER

n this section we extend the material in Sec. 4-5 to more general cases of two ids with different and variable properties.

5-8-1 The Special Case of Pr = Le = 1

e material in Sec. 4-4-2 and 5-4 showed that great simplifications are and direct relations between the velocity and temperature fields result = 1. Similar things happen when also Le = Sc = 1, now involving the on field.

wo species, the energy equation in terms of enthalpy is extended

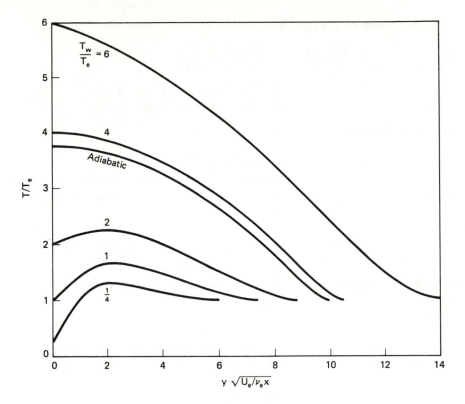

Figure 5-8 Temperature profiles on a flat plate for laminar flow at $M_e = 4$ with various T_w/T_e, $Pr = 0.75$, and the Sutherland viscosity law. (From Van Driest, 1952.)

beyond that in Eq. (5-8) by the extra term shown in Eq. (3-64) to become

$$\left(\rho u \frac{\partial}{\partial x} + \rho v \frac{\partial}{\partial y} \right)\left(h + \frac{u^2}{2} \right)$$

$$= \frac{\partial}{\partial y}\left[\mu \frac{\partial}{\partial y}\left(\frac{h}{Pr} + \frac{u^2}{2} \right) \right] + \frac{\partial}{\partial y}\left[\frac{\mu Le}{Pr}\left(1 - \frac{1}{Le} \right) \sum_i h_i \frac{\partial c_i}{\partial y} \right] \quad (5\text{-}36)$$

The species equation is

$$\rho u \frac{\partial c_i}{\partial x} + \rho v \frac{\partial c_i}{\partial y} = \frac{\partial}{\partial y}\left(\frac{\mu Le}{Pr} \frac{\partial c_i}{\partial y} \right) \quad (5\text{-}37)$$

and the momentum equation remains as the steady form of Eq. (3-25).

Following the logic in Sec. 5-4, we postulate here $c_i = c_i(u)$. With that, the diffusive term on the right-hand side of Eq. (5-37) becomes

$$\frac{\partial}{\partial y}\left(\frac{\mu \mathrm{Le}}{\mathrm{Pr}}\frac{\partial c_i}{\partial y}\right) = \frac{\partial}{\partial u}\left(\frac{\mu \mathrm{Le}}{\mathrm{Pr}}\frac{dc_i}{du}\frac{\partial u}{\partial y}\right)\frac{\partial u}{\partial y}$$

$$= \frac{\mu \mathrm{Le}}{\mathrm{Pr}}\left(\frac{\partial u}{\partial y}\right)^2\frac{d^2 c_i}{du^2} + \frac{\partial}{\partial y}\left(\frac{\mu \mathrm{Le}}{\mathrm{Pr}}\frac{\partial u}{\partial y}\right)\frac{dc_i}{du} \qquad (5\text{-}38)$$

The left-hand side of Eq. (5-37) becomes

$$\frac{dc_i}{du}\left(\rho u \frac{\partial u}{\partial x} + \rho v \frac{\partial u}{\partial y}\right) \qquad (5\text{-}39)$$

Substituting for the term in parentheses from the momentum equation, Eq. (5-37) as a whole then becomes

$$\frac{dc_i}{du}\left(\frac{\partial}{\partial y}\left(\mu \frac{\partial u}{\partial y}\right) - \frac{dp}{dx}\right) = \frac{\mu \mathrm{Le}}{\mathrm{Pr}}\left(\frac{\partial u}{\partial y}\right)^2\frac{d^2 c_i}{du^2} + \frac{\partial}{\partial y}\left(\frac{\mu \mathrm{Le}}{\mathrm{Pr}}\frac{\partial u}{\partial y}\right)\frac{dc_i}{du} \qquad (5\text{-}40)$$

Now, for $\mathrm{Pr} = \mathrm{Le} = 1$ and $dp/dx \equiv 0$, this equation admits to the simple solution

$$c_i = D_1 u + D_2 \qquad (5\text{-}41)$$

where the constants D_1 and D_2 are to be determined from the boundary conditions at the wall and the boundary layer edge. Thus

$$c_i = (c_{ie} - c_{iw})\frac{u}{U_e} + c_{iw} \qquad (5\text{-}42)$$

Taking the derivative at the wall gives

$$\left.\frac{\partial c_i}{\partial y}\right|_w = \frac{c_{ie} - c_{iw}}{U_e}\left.\frac{\partial u}{\partial y}\right|_w \qquad (5\text{-}43)$$

which shows that one *must* have mass transfer at the wall for nonzero skin friction; that is, an impermeable wall case will not admit to this solution.

It is also possible to seek solutions of the form $c_i = c_i(h + u^2/2)$ using the species equation and the energy equation. The result for $\mathrm{Pr} = \mathrm{Le} = 1$ is

$$c_i = D_3\left(h + \frac{u^2}{2}\right) + D_4 \qquad (5\text{-}44)$$

Looking back at Eq. (5-36), one can see that for $\mathrm{Le} = 1$, the energy equation with mass transfer takes the same form as that without mass transfer, Eq. (5-8). Thus, the relations for $h(u)$ developed in Sec. 5-4 hold in such cases also.

The utility of these various generalized Crocco integrals can be illustrated by considering two sample mixing flows shown in Fig. 5-9. For the case in Fig. 5-9(A) with an impermeable wall, we cannot use $c_i(u)$, but from Eq.

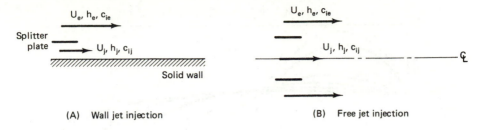

(A)   Wall jet injection   (B)   Free jet injection

Figure 5-9   Schematics of some flows with foreign fluid injection.

(5-44) one finds that

$$\frac{\partial c_i}{\partial y}\bigg|_w = D_3 \frac{\partial h}{\partial y}\bigg|_w \qquad (5\text{-}45)$$

so we can use $c_i(h + u^2/2)$ for an impermeable wall as long as it is also insulated. Applying the initial conditions in the wall jet and the external flow to Eq. (5-44) gives

$$\frac{c_i - c_{ie}}{c_{ij} - c_{ie}} = \frac{(h + u^2/2) - (h + u^2/2)_e}{(h + u^2/2)_j - (h + u^2/2)_e} \qquad (5\text{-}46)$$

On the other hand, we could choose to use the $h(u)$ relation in Eq. (5-19) if $dp/dx = 0$ and the wall is *not* insulated. Applying the conditions in the wall jet and the external flow results in

$$h + \frac{u^2}{2} = \left(\left(h + \frac{u^2}{2}\right)_e - h_w\right)\frac{u}{U_e} + h_w \qquad (5\text{-}47)$$

with the restriction that

$$\frac{(h + u^2/2)_e - h_w}{(h + u^2/2)_j - h_w} = \frac{U_e}{U_j} \qquad (5\text{-}48)$$

For the case in Fig. 5-9(B), it can be easily shown that, with $dp/dx = 0$, Eq. (5-9) applies since $(\partial u/\partial y)_{y=0} = (\partial h/\partial y)_{y=0} = 0$ by symmetry. Using the initial conditions in the jet and the external flow leads to

$$\frac{(h + u^2/2) - (h + u^2/2)_e}{(h + u^2/2)_j - (h + u^2/2)_e} = \frac{u - U_e}{U_j - U_e} \qquad (5\text{-}49)$$

with no restrictions. Since in addition $(\partial c_i/\partial y)_{y=0} = 0$ by symmetry, $c_i(u)$ as in Eq. (5-41) also applies. Using the initial conditions, we get

$$\frac{c_i - c_{ie}}{c_{ij} - c_{ie}} = \frac{u - U_e}{U_j - U_e} \qquad (5\text{-}50)$$

with no restrictions. Thus, in a flow problem as in Fig. 5-9(B), it would only be

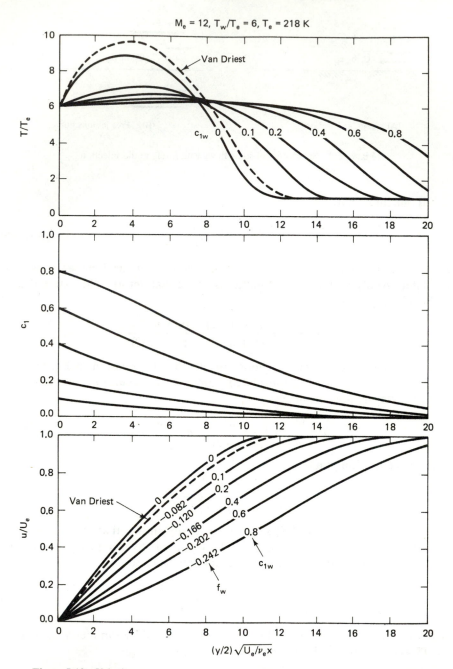

Figure 5-10   Velocity, temperature, and mass fraction profiles for H_2 injection into high-speed airflow over a flat plate. (From Eckert et al., 1958.)

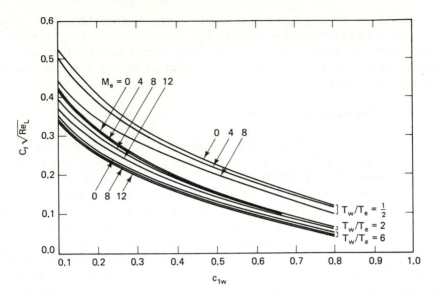

(A)   Skin friction coefficient

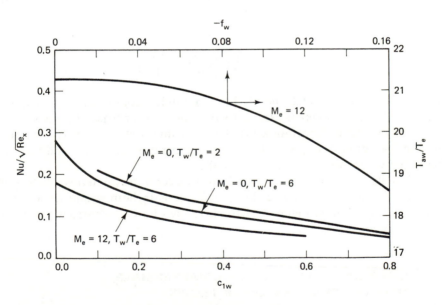

(B)   Dimensionless heat transfer rate

Figure 5-11   Skin friction and heat transfer for H_2 injection into an airstream over a flat plate. (From Eckert et al., 1958.)

necessary to solve the momentum equation for $u(x, y)$ since the enthalpy and species fields can be found directly from Eqs. (5-49) and (5-50) *if* one is willing to accept the restriction Pr = Le = 1. (We note here as an aside that such an approximation is quite reasonable for turbulent jet mixing flows no matter what species are involved.)

5-8-2 Foreign Gas Injection

Exact solutions for injection of a light gas (H_2) through a porous wall into a high-speed airstream without restrictive assumptions on the physical properties and with variable density have been published by Eckert et al. (1958). Such a case is of practical interest for cooling a surface in a high-speed flow. The solutions obtained, however, were restricted to a *similar* distribution of injection (i.e., $v_w \sim x^{-1/2}$), and that is unrealistic from a practical viewpoint. Nonetheless, some useful conclusions can be drawn from the results.

Results for a case treated earlier by Van Driest without injection are shown in Fig. 5-10 for various values of c_{1w} (H_2 concentration at the wall) corresponding to various dimensionless injection rates f_w through a relation such as Eq. (4-42c).

$$\rho_w v_w = \frac{-(\rho D_{12})_w}{1 - c_{1w}} \frac{\partial c_1}{\partial y}\bigg|_w \tag{5-51}$$

where

$$v_w = -\frac{1}{2}\left(\frac{\rho_w}{\rho_e}\right)\left(\frac{U_e v_e}{x}\right)^{1/2} f_w \tag{5-52}$$

The results for $c_{1w} = 0$ (i.e., no H_2 injection) differ from those of Van Driest because of slightly different assumed values of the properties of pure air. The peaks in the static temperature profiles are reduced sharply by H_2 injection even in low amounts ($c_{1w} = 0.1$). The wall temperature gradients, and thus the heat transfer, are also reduced substantially.

Results for skin friction and heat transfer rate (in terms of Nusselt number) are given in Fig. 5-11. Large reductions in both quantities are produced by H_2 injection. Figure 5-11(B) also shows how T_{aw} is strongly reduced by H_2 injection at $M_e = 12$.

5-9 NUMERICAL SOLUTIONS OF COMPRESSIBLE, LAMINAR BOUNDARY LAYER FLOWS

In principle, the numerical solution of compressible, boundary layer problems follows directly from that for incompressible problems. The main differences are the direct coupling with a now nonlinear energy equation and the necessity for calculating the variable physical properties. The evaluation and use of variable physical properties presents no real difficulties within a numerical solution procedure. The properties are evaluated in subroutines using either a table-lookup procedure or curve fits. Since the equations are solved by considering a number of node points across the layer, it is no problem to employ the appropriate variable properties point to point. At high Mach number ($M_e \approx 6$ or greater), one generally has also to deal with nonperfect gas behavior and chemical reactions. Those matters are discussed in a detailed Review by Blottner (1970), but they are beyond the scope of this book.

Virtually all modern methods are based on an implicit finite difference formulation, and the well-developed method of Flügge-Lotz and Blottner (1962) has been selected here as a representative example. This is a *fully implicit* method; that is, the second derivatives with respect to y are evaluated at the downstream station toward which the solution is marching. First, to demonstrate the adequacy of the method, one can compare the results obtained with the finite difference method to the exact results for the relatively

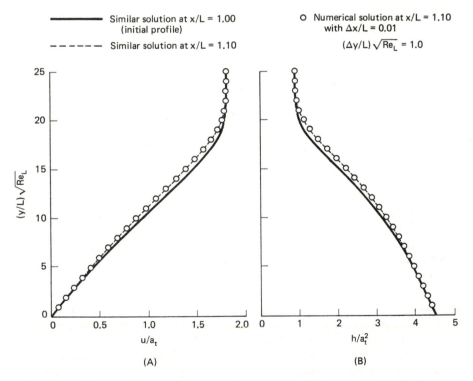

Figure 5-12 Comparison of similar solution and nonsimilar, numerical solution from Flügge-Lotz and Blottner (1962) for laminar flow over a flat plate at Mach 3.0 and $T_w/T_{aw} = 2.0$.

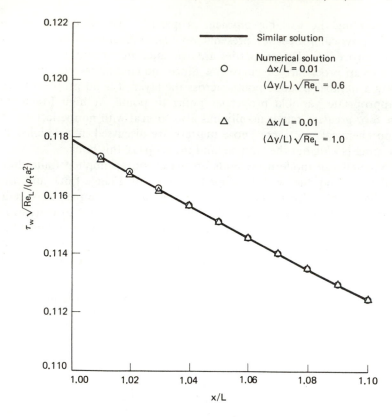

Figure 5-13 Influence of transverse step-size variation in the numerical solution of Flügge-Lotz and Blottner (1962).

simple case of a *similar*, compressible boundary layer flow with $\mu \sim T$ from Low (1955). A *similar* profile is used as the initial condition for the finite difference calculations, and the results after 10 downstream steps are shown in Fig. 5-12. Note that the ordinate here is *not* a similarity variable. Excellent agreement with the appropriate *similar* profile at this downstream station can be seen.

The effects of employing various grid spacings Δx and Δy are displayed in Figs. 5-13 and 5-14. Clearly, good accuracy can be obtained for a case such as this with about 20 points across the boundary layer. Nearly doubling the number of points (decreasing $(\Delta y/L)\sqrt{Re}$ from 1.0 to 0.6) produces no significant changes. Observe the expanded scales on the ordinates. Also, once a suitable Δx has been chosen, reducing it further by a factor of 10 achieves nothing useful.

Anderson and Lewis (1971) have used the Blottner method in Levy–Lees coordinates to treat both laminar and turbulent problems with either the assumption of perfect gas behavior or local chemical equilibrium. A compari-

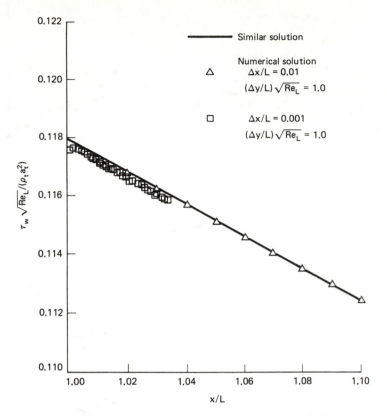

Figure 5-14 Influence of streamwise step-size variation in the numerical solution of Flügge-Lotz and Blottner (1962).

son of prediction and data for a high-Mach-number, high-temperature flow over a flat plate is shown in Fig. 5-15. Here we are concerned only with the laminar regime. It is interesting to note that the difference between the perfect gas and real gas predictions was only 3% at these conditions.

5-10 PRESSURE GRADIENTS AND SEPARATION IN HIGH-SPEED FLOWS

For subsonic, compressible flows below the transonic regime (roughly $M_\infty <$ 0.8 for slender bodies), the general picture of adverse pressure gradients and separation resembles that for incompressible flows as described in Sec. 1-5 and Fig. 1-6. The qualitative influence of favorable pressure gradients is also the same as for low-speed cases. Modern numerical methods for the boundary layer equations are capable of good predictions of these phenomena.

For supersonic flows the flow fields develop differently. First, adverse

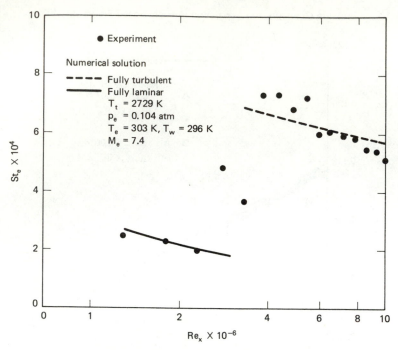

Figure 5-15 Comparison of experiment and numerical solution for laminar high-Mach-number flow over a flat plate. (From Andersen and Lewis, 1971.)

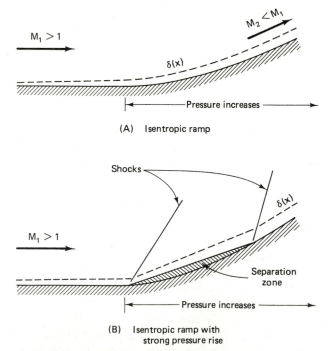

Figure 5-16 Sketches of typical flow fields for laminar boundary layers on supersonic compression surfaces where the inviscid flow is shock-free.

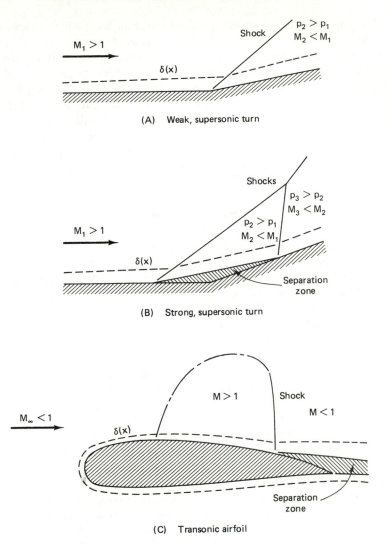

Figure 5-17   Sketches of typical flow fields for laminar boundary layers on high-speed compression surfaces where the inviscid flow contains shocks.

pressure gradients are produced by surfaces turning *into* the flow, as shown in Fig. 5-16(A), rather than *away from* the flow, as depicted in Fig. 1-6. The flow field in Fig. 5-16(A) is generally called an *isentropic ramp*, since the goal is to produce a compression without shocks in the flow. If the pressure gradient is too strong, separation will occur as shown in Fig. 5-16(B). Boundary layer analysis is capable of good predictions up to separation. The shocks are produced by the separation; the interaction of the boundary layer with the shocks cannot generally be analyzed with boundary layer theory.

It is easy to envision supersonic flows over surfaces where shocks will be produced even with an inviscid analysis. This is shown in Fig. 5-17(A). For subsonic approach flows near Mach 1.0, the presence of a body usually accelerates the flow to locally supersonic speeds. As the flow decelerates back down toward the subsonic freestream value, a shock is generally formed which impinges on the body boundary layer. For all but the weakest shocks, in either case, separation of a laminar boundary layer will result, as shown in Fig. 5-17(B) and (C). In these cases, the shocks separate the boundary layer as a result of a large, almost discontinuous, pressure rise, but again this type of interaction cannot be analyzed with the boundary layer theory.

PROBLEMS

5.1. Plot $T_{kinetic}/T_t$ versus V for air and CO_2 up to 3000 m/s.

5.2. Consider the flow of H_2 over a flat plate at $M_\infty = 2.0$ at $p_\infty = 10^{-1}$ atm and $T_{t,\infty} = 600$ K. What is the adiabatic wall temperature? If the plate is 20 cm long and 10 cm wide, how much cooling is necessary to maintain the plate at 300 K?

5.3. Prove that for $Pr = 1$, the maximum static temperature in the boundary layer over a flat plate occurs where u is given by Eq. (5-24). (*Hint*: Differentiate with respect to u.)

5.4. Plot $T(u)$ for the numerical example in Sec. 5-4.

5.5. Find the adiabatic wall temperature for a surface with a local Mach number of 0.85 at an altitude of 10,000 m, assuming a *standard* atmosphere.

5.6. Find u/U_e versus y for a flat-plate boundary layer in air with $M_\infty = 3.0$, $p_\infty = 10^{-1}$ atm, $T_{t,\infty} = 300$ K, and $T_w/T_\infty = 1.0$ where $\delta = 1$ cm. (*Hint*: Use the Crocco integral, the Howarth transformation, and the Blasius solution.)

5.7. Plot δ and δ_T versus M_e for flat-plate flow of air with $Re_L = 10^5$ with an adiabatic wall and with $T_w/T_\infty = 1.0$.

5.8. Rederive the Crocco integrals for $h(u)$ for unsteady cases.

5.9. Consider a porous flat plate in a hypersonic wind tunnel at $M = 12.0$ with $p_t = 30$ atm and $T_e = 200$ K. If the wall temperature is 300 K, what is the percentage reduction in heat transfer rate at a station where $Re_x = 10^5$ for a case with H_2 injection sufficient to give $c_{1w} = 0.2$ compared to a case with no injection?

5.10. For the conditions of Prob. 5.9, what is the rate of H_2 injection required?

6

TRANSITION
TO TURBULENT FLOW

6-1 INTRODUCTION

The material in Chapters 1 to 5 completed the study of laminar boundary layer flows at the level intended for this book. Chapters 7 to 9 are concerned with turbulent boundary layer flows. In this chapter, one of the most important and probably the most difficult subject in boundary layer work is addressed—the prediction of transition from laminar to turbulent flow. In Chapter 1 some features of turbulent flows were introduced, and it has been stressed that most flows of engineering interest are turbulent. But how is the fluid dynamicist to be certain that a flow will be laminar or turbulent under a given set of conditions? This question is of more than casual interest, because the levels of skin friction and heat transfer, for example, are quite different in the two cases. Common experience tells us that high-Reynolds-number flows are more likely to be turbulent, but such a crude statement is hardly satisfactory for careful design. The designer needs accurate methods for predicting transition, but unfortunately these are hard to achieve. In Secs. 6-2 to 6-4, attempts at the analytical prediction of transition and their limitations are discussed. The remainder of the chapter is concerned with selected empirical information on the subject. This coverage is an accurate reflection of the current state of the art in this area, where much more trust is placed in empiricism than in analysis.

Before examining some of the analytical prediction techniques, it is informative to study briefly the earliest organized experimental investigations of transition-Reynolds (1883) classical pipe flow experiments. This can be supple-

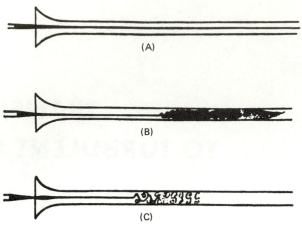

(A)–(C) Reynolds's sketches of flow observation:
(A) low speed (low Re_D); (B) higher speed
(higher Re_D); (C) spark illumination

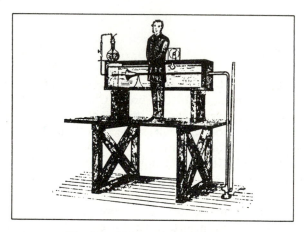

(D)   Reynolds's experimental apparatus

Figure 6-1   The classic pipe flow experiments of Reynolds (1883).

mented by the reader's own observations of transition phenomena in a water faucet efflux as the valve is opened (increasing the Reynolds number) or in the smoke plume from a cigarette resting in an ashtray as two common examples. Reynolds' experiments were conducted in an apparatus as shown schematically in Fig. 6-1. As the flow rate through the glass pipe was increased, the behavior of the dye stream changed from that shown in Fig. 6-1(A) to that in Fig. 6-1(B), indicating a transition from laminar to turbulent flow. By increasing the velocity in a fixed size pipe D with the same fluid (constant ρ and μ),

Reynolds was increasing the dimensionless grouping, $\rho V D / \mu \equiv \mathrm{Re}_D$, named after him. He found that the value for transition, called the *critical* Reynolds number, was approximately 2300. Later studies have shown that this critical value is very sensitive to the inlet conditions. By careful tailoring of the inlet flow, it has been possible to maintain laminar flow up to $\mathrm{Re}_D \approx 20{,}000$. A value of 2000 is commonly used in engineering practice, but this extreme sensitivity to the conditions of the experiments persists throughout all transition work. It contributes to the apparent *scatter* in the empirical information, but actually one is seeing the effects of small but important differences in seemingly identical experiments.

6-2 HYDRODYNAMIC STABILITY THEORY

Visual observations of turbulence clearly indicate that the transition process is the end result of the growth of initially small, probably random disturbances in the flow. Small disturbances due to, for example, noise or slight vibrations of solid surfaces are always present in the background of any flow. Apparently, under some conditions in the flow, these disturbances are damped out, whereas at other conditions, they are amplified. From Reynolds' experiments, one can expect that the value of a suitable Reynolds number will describe the boundary between the two cases. This situation bears a close resemblance to the vibration of mechanical systems, which had been analyzed rather thoroughly by the early 1900s. Thus, it should not be surprising that attempts were made to apply the same general methods to the question of the stability of a flow to small disturbances. These attempts were quite successful and yet unsuccessful at the same time (this will become clear shortly). Also, the mathematics of the analysis proved very complex and interesting (to mathematicians). For these reasons, we give only a brief overview of the theory here. For the interested reader, there are whole books devoted to the subject.

In general, one wishes to analyze the unsteady, three-dimensional behavior of a small disturbance in a baseline laminar boundary layer flow. This is obviously a difficult undertaking, and some simplifying assumptions are required. First, we get a helpful assist from a theorem by Squire (1933) which proved that two-dimensional disturbances are always less stable than three-dimensional disturbances. Thus, one need only consider the two-dimensional case to find the minimum instability conditions. The principal simplifying assumptions are: (1) the disturbance quantities remain small compared to the baseline flow (if $u = u_0 + \hat{u}$, $\hat{u} \ll u_0$, etc.), (2) the baseline flow is a function only of the transverse coordinate [i.e., $u_0 = u_0(y)$ alone, etc.], and (3) the disturbance can be written in terms of a stream function as

$$\hat{\psi}(x, y, t) = \phi(y) \exp\left[i\alpha(x - ct)\right] \tag{6-1}$$

Here $i \equiv \sqrt{-1}$, $\phi(y)$ is a complex amplitude function, α is the wavenumber of the disturbance ($= 2\pi/$wavelength), and c is the complex phase velocity ($= c_r + ic_i$). Looking at Eq. (6-1), one can see that stability with respect to time depends on the value of αc_i. For $\alpha c_i < 0$, there is damping (i.e., the disturbance decreases with time); for $\alpha c_i = 0$, a neutral condition is achieved and for $\alpha c_i > 0$, amplification or instability results. Perhaps it is helpful to note that the form assumed in Eq. (6-1) implies that

$$\hat{u} = \frac{\partial \hat{\psi}}{\partial y} = \frac{\partial \phi}{\partial y} \exp \left[i\alpha(x - ct) \right]$$

$$\hat{v} = -\frac{\partial \hat{\psi}}{\partial x} = -i\alpha\phi(y) \exp \left[i\alpha(x - ct) \right] \tag{6-2}$$

With all of this, the full, viscous equations of motion for an incompressible, constant-property flow can be simplified to

$$(u_0 - c)(\phi'' - \alpha^2 \phi) - u_0'' \phi = \frac{-i}{\alpha \text{Re}} (\phi^{(iv)} - 2\alpha^2 \phi'' + \alpha^4 \phi) \tag{6-3}$$

This is the famous Orr–Sommerfeld equation, and it proved difficult to solve for general profile shapes $u_0(y)$ before the advent of the large digital computer. For that reason, much of the earliest work concentrated on cases with a linear profile ($u_0 \sim y$, i.e., $u_0'' \equiv 0$) and/or inviscid flow (Re $\rightarrow \infty$).

Various calculations for the general case have been presented. Lin (1945) and others used asymptotic theory, and in more recent times, direct numerical solutions have been obtained such as those by Wazzan et al. (1968). These results are shown in Figs. 6-2 and 6-3. The results of the theory have been

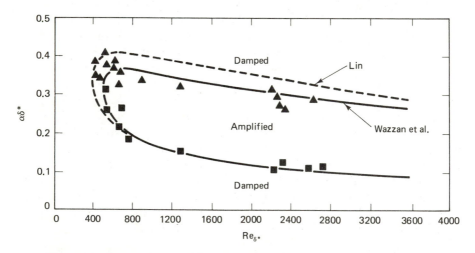

Figure 6-2 Comparison of the predictions of the hydrodynamic stability theory by Lin (1945) and Wazzan et al. (1968) for the wavelength of neutral disturbances for flow over a flat plate with the measurements of Schubauer and Skramstad (1947).

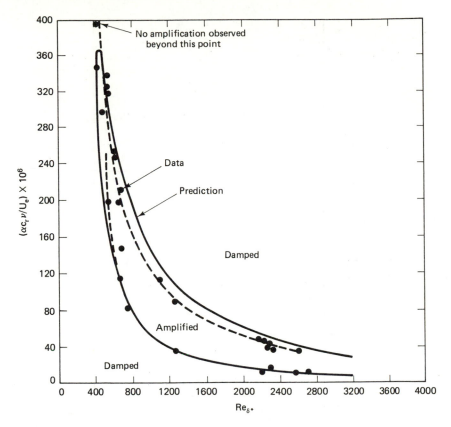

Figure 6-3 Comparison of the prediction of the hydrodynamic stability theory by Lin (1945) for the frequency of neutral disturbances for flow over a flat plate with the measurements of Schubauer and Skramstad (1947).

confirmed by experiment as far as the onset of instability is concerned. Schubauer and Skramstad (1947) placed a thin magnetic ribbon in a laminar, flat-plate boundary layer and then excited the ribbon externally to vibrate, producing disturbances at various, known wavelengths. Theory and experiment agreed rather well, as shown in the figures.

It can be seen that the maximum Reynolds number (based here on δ^*) where a disturbance is stable depends on the wavenumber α of the disturbance, but that is generally unknown for background noise. Therefore, the only rational choice is to focus on the worst case which gives for this problem $(Re_{\delta*})_{crit} = 520$ for $\alpha\delta^* \approx 0.30$. It is informative to observe that the wavelengths of typical unstable disturbances are large compared to the boundary layer thickness. The smallest unstable wavelength corresponds to $\alpha\delta^* \approx 0.30$, which implies a minimum wavelength of $2\pi\delta^*/0.30 \approx 21\,\delta^* \approx 7\delta$.

If the onset of instability is predicted well, does that mean that transition

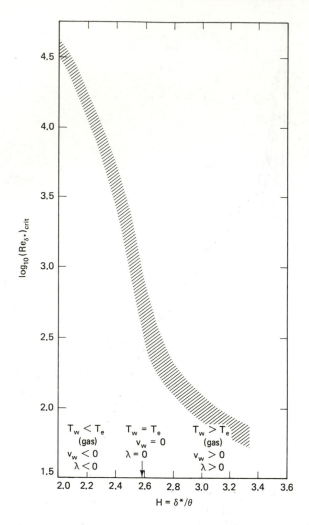

Figure 6-4 Minimum Reynolds number for instability in low-speed flow of gases (heating/cooling effects) as influenced by pressure gradient, injection/suction, and heating/cooling. (From Stuart, 1963.)

is also predicted well? Unfortunately, the answer is *no*. Experiment indicates that transition on a flat plate in a relatively *clean* external flow (a low free-stream turbulence level of about 0.1%) occurs at about $(Re_x)_{trans} \approx 2.5 \times 10^6$. Using the Blasius solution, this translates to $(Re_{\delta*})_{trans} \approx 2700$, which is much higher than the instability condition at 520. The difficulty is simply that instability and transition are not the same thing. Instability is merely the very early precursor of transition, and knowing the conditions for instability is not of much use in predicting the subsequent conditions for transition. We have discussed here only the simple flat-plate boundary layer case, but the pessimistic conclusions reached on the basis of that single example are general.

The hydrodynamic stability theory is useful in indicating which conditions will tend to hasten or delay transition in a relative sense. For example, one finds that profiles that have an inflection point are very unstable. Since an adverse pressure gradient leads toward profiles with an inflection point, an adverse pressure gradient will hasten the onset of instability and transition. These qualitative predictions are confirmed by experiment.

Distributed, transverse injection, as through a porous surface, will also produce profiles with an inflection point. So can surface heating. This can be seen by examining the momentum equation for a flat-plate flow ($dp/dx = 0$) evaluated at the surface ($u = v = 0$), which then becomes

$$0 = \frac{\partial}{\partial y}\left(\mu(T)\frac{\partial u}{\partial y}\right) \tag{6-4}$$

The curvature of the profile at the surface may be written

$$\frac{\partial^2 u}{\partial y^2} = -\frac{1}{\mu}\left(\frac{\partial \mu}{\partial y}\right)\left(\frac{\partial u}{\partial y}\right) \tag{6-5}$$

For a heated surface, the temperature decreases with y. For gases, the viscosity increases with temperature, and $\partial\mu/\partial y < 0$, leading to $(\partial^2 u/\partial y^2)_w > 0$. Near the outer edge of the layer, the curvature is always negative, so surface heating in a gas flow implies an inflection point in the layer and decreased stability. Since for liquids the viscosity generally decreases with temperature, this whole picture is reversed.

All of these influences on profile shape can be expressed in terms of the shape parameter, $H \equiv \delta^*/\theta$. The effect of variations in H on stability is shown in Fig. 6-4 from Stuart (1963) for gases. The effects of pressure gradient and suction/injection are the same for gases and liquids, but the effects of heating/cooling are reversed.

6-3 THE e^{10} METHOD

For many years, the hydrodynamic stability theory rested largely unused by working fluid dynamicists, because it could not be used to predict *transition* for a given flow problem. Recently, a clever heuristic method for extending the stability theory to produce approximate predictions of transition has been developed by a group at Douglas Aircraft Co. [see Jaffe et al. (1970) for details]. The basic idea is to track the amplification rate of a disturbance from the point of neutral stability downstream along the surface until the integrated value with surface distance reaches a certain, hopefully universal, value. Since the amplification rate grows rapidly once the flow is past the neutral stability point (zero amplification rate), this idea is very appealing from a physical viewpoint.

There are three steps in the development of this method. First, one must deal with the *spatial* behavior of *small disturbances*, not the *temporal* or *timewise* behavior treated before. The two are identical only at the neutral stability point. The derivation of the spatial analysis is not, however, difficult. In Eq. (6-1), α becomes a complex number, the imaginary part of which, α_i, becomes the amplification rate. Equation (6-3) with suitable boundary conditions has again to be solved.

The second step is the generation of laminar velocity profiles at various stations along the body surface of interest. These are to be used as the baseline profiles $u_0(y)$ for stability calculations at a number of streamwise stations from the leading edge (or front stagnation point) to a point well past the streamwise location where transition may be expected based on some cruder estimation technique or data correlations. Obviously, accurate profiles are required or else the stability calculations based on them will be in error. Also, since calculations at a number of stations must be made, efficiency is important. A modern, implicit finite difference procedure such as that of Blottner (1970) can be recommended for this purpose.

The third step is to implement the method and see if comparisons with experiment yield any universal value for the integrated amplification rate as indicative of transition. The integrated amplification rate is determined in

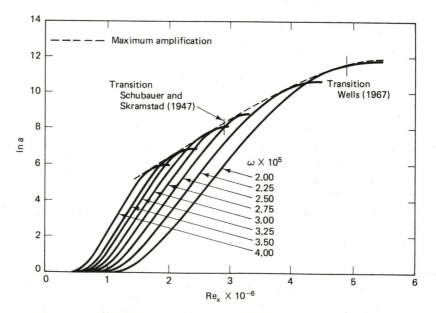

Figure 6-5 Calculated development of amplification factors along a flat plate. (From Jaffe et al., 1970.)

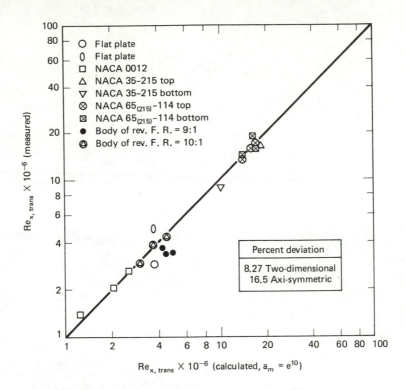

Figure 6-6 Comparison of transition prediction by the e^{10} method and experiment. (From Jaffe et al., 1970.)

terms of an amplification factor a defined as

$$a(\bar{x}) = \exp\left[-\frac{U_\infty L}{v} \int_{\bar{x}_n}^{\bar{x}} \frac{\alpha_i}{U_e \delta/v} \frac{U_e}{U_\infty} d\bar{x} \right] \qquad (6\text{-}6)$$

where $\bar{x} = x/L$ and $\bar{x}_n$ denotes the location of the neutral stability point. The frequency is made dimensionless as ω by multiplying the dimensional frequency by v/U_e^2. The value of $a(\bar{x})$ at any streamwise location will vary with frequency and exhibit a maximum $a_m(\bar{x})$.

Figure 6-5 shows the variation of the logarithm of $a(\bar{x})$ versus Re_x for flow over a flat plate with dimensionless disturbance frequencies from 2×10^{-5} to 4×10^{-5}. The envelope of these curves corresponds to the maximum amplification factor $a_m(\bar{x})$. The high and low values of $(Re_x)_{trans}$ reported in the literature for low-free-stream-turbulence tests are shown, and these correspond to $\ln(a_m(\bar{x})) = 8.3$ and 11.8.

Extensive comparisons with experiments for flows over a variety of bodies lead to an average value of $\ln(a_m(\bar{x})) = 10$ to indicate transition. The results of applying this rule compared to experiment are shown in Fig. 6-6.

(A) $Re_L = 20,000$; Stable.

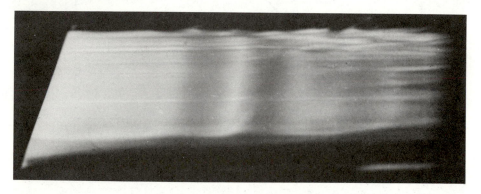

(B) $Re_L = 100,00$; Disturbances grow.

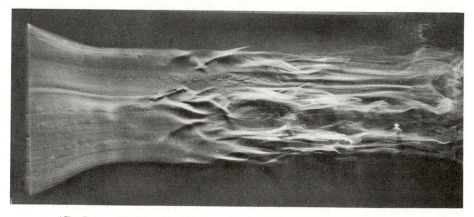

(C) $Re_L = 100,000$ with 1 deg. angle of attack to produce larger disturbances; Transition.

Figure 6-7 Photographs of dye streaks in a boundary layer on a flat plate in a water flow showing the growth of disturbances leading to transition. (From Werle, 1982 at ONERA.)

Although the precision achievable is not great, a workable, if complicated method based on rather sound concepts for predicting transition has clearly been developed. It can be extended to include the effects of heat transfer, compressibility, and so on.

6-4 A METHOD BASED ON Re_θ

A cruder but very simple method for roughly predicting transition was presented by Michel (1952). He succeeded in correlating transition location for low-speed flows on the basis of only the Reynolds number based on momentum thickness, $Re_\theta \equiv \rho U_e \theta / \mu$ as

$$Re_{\theta,\,\text{trans}} \approx 2.9 Re_{x,\,\text{trans}}^{0.4} \qquad (6\text{-}7)$$

where $Re_x \equiv \rho U_e x / \mu$. The required distribution of $\theta(x)$ can be calculated by the Thwaites–Walz method (see Sec. 2-3-2).

6-5 SELECTED EMPIRICAL INFORMATION

6-5-1 The Nature of Transition

The first thing that is important to understand about transition is that it is not a single, abrupt event in the flow, but rather a process involving several steps that occur over a region of the flow. The steps involved can be seen in the excellent, detailed photographs of Werle (1982) shown in Fig. 6-7. After the

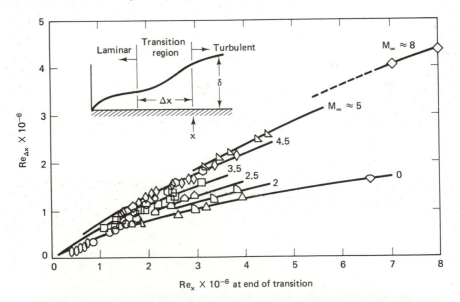

Figure 6-8 Correlation of the length of the transition region with the Reynolds number for transition and Mach number. (From Potter and Whitfield, 1962.)

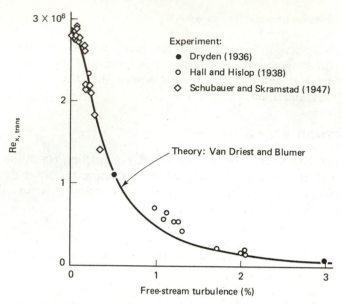

Figure 6-9 Comparison of prediction and experiment for the effect of free-stream turbulence on transition for flow over a flat plate. (From Van Driest and Blumer, 1963.)

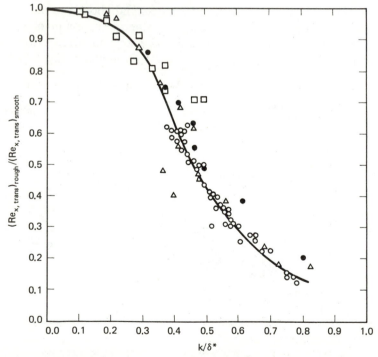

Figure 6-10 Ratio of the transition Reynolds number for flow over a flat plate with a single roughness element to that for a clean plate. (Collected from various sources by Dryden, 1953.)

neutral stability point, two-dimensional disturbances begin to grow. These then break down into three-dimensional disturbances which themselves break down. Soon localized *spots* of turbulence appear, and, finally, these merge together to produce fully turbulent flow. The length of the transition region Δx_{tr} as a function of Mach number and the Reynolds number at transition has been correlated by Potter and Whitfield (1962) as shown in Fig. 6-8.

6-5-2 Free-Stream Turbulence

Intuitively, one might expect that the level and nature of the background disturbances present in the flow would influence transition, and that is indeed the case. Some data for low-speed flow over a flat plate are given in Fig. 6-9 together with the results of a theory by Van Driest and Blumer (1963). This theory is based on the simple, physical notion that transition will correspond to a critical value of a Reynolds number based on the vorticity in the boundary layer

$$\text{Re}_{\text{vort}} \equiv \frac{(\partial u/\partial y)y^2}{\nu} \tag{6-8}$$

For laminar velocity profiles described by the Pohlhausen pressure gradient parameter λ [see Eqs. (2-24) and (2-26)], it can be shown that

$$\frac{(\text{Re}_{\text{vort}})_{\text{trans}}}{\text{Re}_\delta} = A + B\lambda + C \cdot \text{Re}_\delta \left(\frac{\overline{(u')^2}}{U_e^2} \right) \tag{6-9}$$

where A, B, and C are constants to be found by comparison with experiment. Van Driest and Blumer found that

$$\frac{9860}{\text{Re}_\delta} = 1.0 - 0.049\lambda + 3.36\text{Re}_\delta \left(\frac{\overline{(u')^2}}{U_e^2} \right) \tag{6-10}$$

6-5-3 Roughness

The description of surface roughness is one of those things that sounds simple but really is not. The character of the roughness is obviously different for a machined surface than for a corroded surface, although corroded steel and aluminum surfaces are also quite different from each other. All of this has led to considerable difficulty in describing the influence of roughness in terms of any single parameter, such as the average roughness height k. Most careful studies have been done for a single, isolated, two-dimensional roughness element—usually a wire on the surface, or for the simple distributed roughness produced by uniform-size sand glued to the surface.

Dryden (1953) has collected data for the influence of single, two-dimensional roughness as plotted in Fig. 6-10 in terms of $(\text{Re}_{x,\,tr})_{\text{rough}}/(\text{Re}_{x,\,tr})_{\text{smooth}}$.

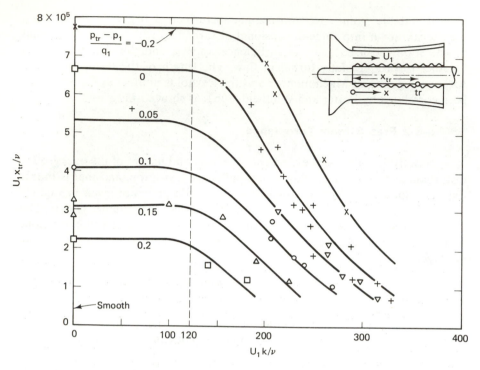

Figure 6-11 Effect of uniform sand grain roughness and pressure gradient on transition for low-speed flow. (From Feindt, 1956.)

Feindt (1956) studied the combined effects of distributed, uniform-size-sand roughness and pressure gradient in the device shown in the insert in Fig. 6-11. Here we are concerned only with the effects of roughness alone, so focus on the curve labeled "0". Clearly, for very small k, there is no influence. However, for $U_1 k/\nu \geq 120$, a rapidly growing effect appears. This value may be used to define a so-called *critical roughness size*.

6-5-4 Bluff Bodies at Low Speeds

The effects of transition on the flow over a bluff body such as a sphere or a circular cylinder at low speeds are sufficiently dramatic so as to warrant a separate discussion. Consider first the drag coefficient of a sphere versus Reynolds number as plotted in Fig. 6-12. A rapid drop in the drag occurs at about $C_D = 0.3$, where $\mathrm{Re}_D \approx 3.5 \times 10^5$. Flow visualization studies show that this drag decrease occurs at conditions where transition in the boundary layer on the sphere takes place before the laminar separation point (82°). Laminar separation can be clearly seen in Fig. 1-8(A). We will see later that a turbulent boundary layer can sustain a larger adverse pressure gradient before separation than a laminar one. This means that the flow remains attached further

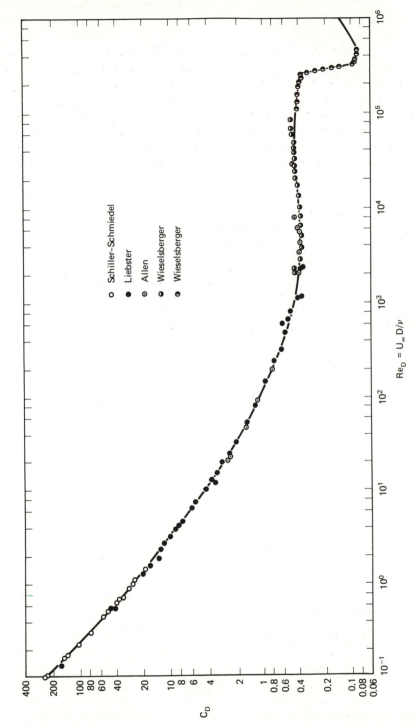

Figure 6-12 Experimental data for the drag coefficient of a sphere in low-speed flow as a function of Reynolds number from various workers.

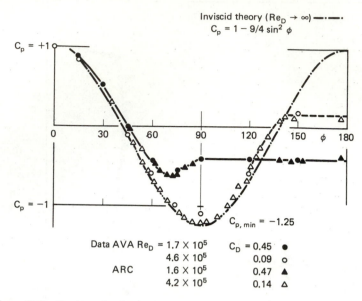

Figure 6-13 Pressure distribution around a sphere in low-speed flow for various Re_D.

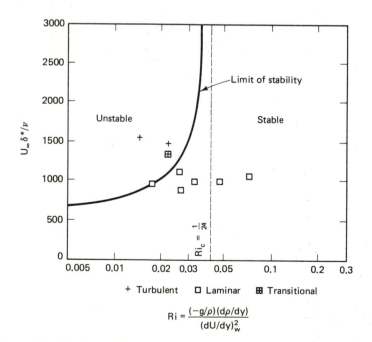

$$Ri = \frac{(-g/\rho)(d\rho/dy)}{(dU/dy)_w^2}$$

Figure 6-14 Density-stratified flow over a flat plate as a function of Richardson number. (From Schlichting, 1935.)

along the surface of the sphere, keeping the pressure distribution close to the inviscid case (drag = 0) longer. This is displayed in Fig. 6-13. Thus, even though the skin friction drag is higher in the turbulent case, the pressure drag is reduced sharply resulting in a lower total drag.

6-5-5 Density-Stratified Flows

The stability of a column of fluid whose density is varying due to temperature and/or pressure variations is important in atmospheric and ocean flows. The key dimensionless grouping is called the Richardson number:

$$Ri \equiv \frac{-g(d\rho/dy)}{\rho(du/dy)_w^2} \tag{6-11}$$

The term $-g(d\rho/dy)$ is the restoring force on a unit volume of fluid displaced a

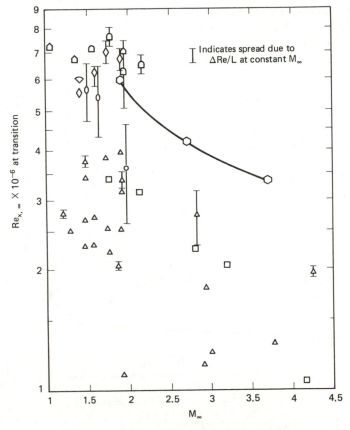

Figure 6-15 Experimental data for the effect of Mach number on transition on sharp, insulated cones. (Data from many sources collected by Wilson, 1966.)

distance y in a density gradient, and the term $\rho(du/dy)^2$ represents the inertia of the fluid with respect to its undisturbed condition. The Richardson number is thus the ratio of the restoring force to the inertia force, and a high Richardson number will indicate a stable condition. Schlichting (1935) studied this situation analytically and found that the Reynolds number for instability tended to infinity for $Ri = \frac{1}{24}$. Experiment by Prandtl and Reichardt (1934) showed that the flow is laminar in the predicted stable zone and turbulent in the predicted unstable zone as shown in Fig. 6-14.

6-5-6 Supersonic Flows

For supersonic flows, the effect of increasing the Mach number is to decrease the transition Reynolds number at least up to $M_\infty \approx 4$. This is shown in Fig. 6-15 for flow over insulated cones. The data shown as squares and triangles may have been influenced by flow irregularities and higher free-stream turbulence. The general effect of T_w/T_e is shown in Fig. 6-16 at Mach 3.0. There has been considerable debate in the literature about the effect of T_w/T_e at high Mach numbers, and the situation is not yet sufficiently clarified to allow discussion here.

The effect of isolated, two-dimensional roughness is shown in Fig. 6-17

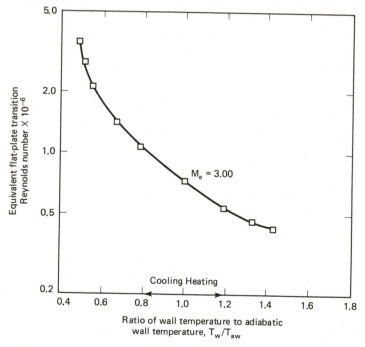

Figure 6-16 Effect of wall temperature on transition at Mach 3.0. Data from cones converted to equivalent flat plate. (From Jack and Diaconis, 1955.)

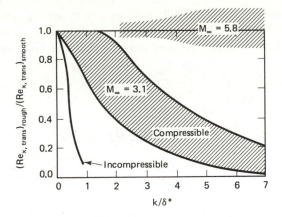

Figure 6-17 Ratio of the transition Reynolds number for flow over a flat plate with a single roughness element to that for a clean plate. ($M = 3.1$ from Brinich, 1954; $M = 5.8$ from Korkegi, 1956.)

for two Mach numbers. For high Mach numbers, it is virtually impossible to induce transition by *tripping* with a wire. The predicted effect of distributed, uniform roughness is indicated in Fig. 6-18 as a function of T_w/T_e at one moderate Mach number. These results show that cooling a rough surface reduces the Reynolds number (based on free-stream conditions) for transition. This is explained by studying more closely the flow near a roughness element on the surface. A cooler fluid implies a lower kinematic viscosity and hence a higher unit Reynolds number based on conditions in that region of the flow.

It is interesting to mention here the phenomenon of *relaminarization*. In high-speed flows, it is possible to produce severe pressure gradients. For exam-

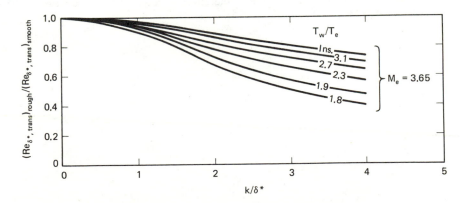

Figure 6-18 Predicted effect of uniform distributed roughness on the transition Reynolds number for flow over a cone at supersonic speed. (From Van Driest and Boison, 1957.)

ple, a rapid-expansion nozzle will produce a very strong, favorable pressure gradient. This can actually cause a turbulent boundary layer to revert to a laminar state. Back et al. (1969) found that this occurred for flows with

$$\frac{v_e}{U_e^2} \frac{dU_e}{dx} > 2 \times 10^{-6} \qquad (6\text{-}12)$$

PROBLEMS

6.1. Consider CO_2 flowing at 2.0 atm and 400 K in a 5.0-cm-diameter pipe in a chemical plant. At what flow rate (in m^3/s) will the flow become turbulent? What if the fluid were H_2?

6.2. For flow over a flat plate, what is the Reynolds number based on x for instability if the disturbance has a wavelength $= 10\delta$?

6.3. What is the transition Reynolds number based on x for flow over an NACA 0012 airfoil? What is the length of transition Δx expressed in terms of a Reynolds number?

6.4. Compare the prediction of the method of Michel with that of Van Driest and Blumer for the transition Reynolds number for flow over a flat plate.

6.5. For water (at 300 K) flow over a flat plate at 3 m/s, compare the length to transition for free-stream turbulence levels of 2.0% and 0.1%. What if the surface had a wire with $k/\delta^* = 0.2$?

6.6. Air at standard temperature and pressure is flowing past a 25-cm-diameter ball on the top of a flag pole. What is the velocity for the drag decrease due to transition? How much will the pressure at the rear of the sphere change?

6.7. For the ARDC Model Atmosphere, what value of the velocity gradient corresponds to the stability limit at 1000 m?

6.8. Compare the transition Reynolds number for flow over an insulated cone at Mach 1.5, 3.0, 4.5, and 6.0. What would be the effect of roughness on the surface with $k/\delta^* = 2.0$ at Mach 3.0 and 6.0?

7

WALL-BOUNDED, INCOMPRESSIBLE TURBULENT FLOWS

7-1 INTRODUCTION AND SCOPE

In this chapter we begin the study of the most important part of boundary layer theory from a practical point of view—turbulent flows. The general subject is large and complex, so it is prudent to start with a portion that one can grasp and then build on. First, we restrict this chapter to constant-density, constant-property (laminar thermophysical properties) flows. Second, only *wall-bounded turbulent flows* are considered. These are flows with a rigid (not necessarily impermeable) surface boundary on at least one side of the turbulent, viscous region. Flows without such a boundary are called *free turbulent flows*, and they are the subject of the next chapter. The reason for this important distinction will become clear shortly.

7-2 ENGINEERING REQUIREMENTS OF TURBULENT ANALYSES

It was indicated in Chapter 1 that turbulent flow is necessarily unsteady and three-dimensional. Moreover, the frequencies of the unsteadiness and the size of the scales of the motion span several orders of magnitude. Some reflection on these basic facts will yield a perspective on the magnitude of the problem of attempting to analyze all the details of a turbulent flow, even a very simple turbulent flow such as that over a flat plate. One example of a specific aspect of this general situation may be helpful. If the scales of the turbulent motion

139

span orders of magnitude in size, how are the grid spacings (Δx, Δy) in a numerical method to be chosen intelligently? Obviously, great crudity will result if the grid spacings are larger than the scales of a significant part of the motion. It is safe to conclude, then, that the task of producing a general analysis for all the details of a turbulent flow is hopeless at the present time. It is unlikely that the picture will change dramatically any time soon.

With the bleak assessment outlined above, is the designer to be left completely without analytical tools? The answer is *no*, and the opening that can be exploited comes from the limited design information required from a turbulent analysis. Referring back to the logic first presented in Chapter 2, we can state that the first requirement of a boundary layer analysis is a prediction of $C_f(x)$, including any points where $C_f = 0$ (i.e., separation). Second, one may be interested in $\delta(x)$ [and perhaps $\delta^*(x)$ and $\theta(x)$]. Other quantities, such as detailed velocity and temperature profiles, are of rapidly diminishing utility for design, in most cases.

Before proceeding further, one must pause here and ask what $C_f(x)$, for example, means in a turbulent flow. For a pipe flow problem with a constant throughput in a chemical plant or a frictional resistance prediction for a ship moving at constant speed, do we need, or want, the instantaneous, fluctuating value of the wall shear at a point on the surface? Obviously not. One wants to know the steady, overall pumping power required to drive the fluid through the pipe or the steady, power plant size and propeller performance required to propel the ship. Even if the pipe throughput or the ship speed were not constant, one would not usually need the fluctuating value of $C_f(x)$. The designer would need the unsteady variation of $C_f(x)$ that is correlated with the unsteady changes in the initial and/or boundary conditions (reread the end of Sec. 1-6 and rethink the distinction between what is called a *steady turbulent flow* and an *unsteady turbulent flow*). The needs of the analyst may be seen as most commonly corresponding to the *time-averaged* or *mean* quantities [e.g., $U(x, y)$] introduced in Sec. 1-6. Such quantities can be steady or unsteady and two-dimensional or three-dimensional (only the fluctuating part of the turbulent motion is always three-dimensional). Also, it is only the size of the scales of the fluctuating part of the motion that spans orders of magnitude.

Taking all of this together, it can be seen that a turbulent analysis aimed only at predicting the mean flow [$C_f(x)$, $\delta(x)$, etc., and perhaps $U(x, y)$, etc.] might well serve the main purposes of the designer and might also be within the range of modern analytical and/or numerical capability. That notion has been taken as defining the scope of this book. Lest the reader relax at this point, it should be noted that only limited success at achieving reliable, efficient predictions of turbulent flows has been attained to date and only then with the expenditure of great amounts of ingenuity and intellectual energy.

It is important to observe that there are some practical situations where a knowledge of the detailed fluctuating character of the motion is essential. An example is ignition and combustion of mixing streams of a fuel and oxidant (e.g., H_2 and air). The rate of conversion of chemical to thermal energy in the flame depends on the instantaneous, local temperature in a very nonlinear fashion. Thus, in order to predict the steady heat release at a point in the flame, one cannot base the rates on $\bar{T}(x, y)$. It would be necessary to calculate the rates with $T'(x, y, z, t)$ and then take the time average of the results. It is also not sufficient to use a quantity such as $\sqrt{(T')^2}$ because of the fact that the *average of a quantity raised to a power is not the same as the power of the average of the quantity*. This type of situation is at the forefront of research in turbulence, and a more detailed discussion is outside the scope of this book.

Before beginning the development of mean flow analyses, it is helpful to consider some of the available empirical information. That will serve as a solid basis for the analytical development.

7-3 EMPIRICAL INFORMATION ON THE MEAN FLOW AS A BASIS FOR ANALYSIS

7-3-1 Flow over a Flat Plate

In a complex field, it is wise to begin with the simplest physical situation that is at least representative of the more general cases. For boundary layers, that is the flow over a flat plate ($dP/dx = 0$). If one compares measurements for the velocity profiles over a flat plate in laminar (see Fig. 4-2) and turbulent cases (see U/U_∞ in Fig. 1-10), a striking difference is apparent, as illustrated in Fig. 7-1. When plotted against y/δ, the mean turbulent velocity profile *appears* to intersect the wall ($y = 0$) at a value greater than $U = 0$. Actually, it proceeds down toward $y/\delta = 0$ decaying slowly and then drops quickly to $U = 0$ at $y/\delta = 0$ over a very small range of y/δ. This behavior can be likened to that which would result for a laminar boundary layer composed of two layers of two different fluids: a high-viscosity fluid in the outer region of about $0.05 \le y/\delta \le 1.00$ and a much lower viscosity fluid near the wall of about $0 \le y/\delta \le 0.05$. Since a simple force balance will show that the shear at the interface must be continuous, we can write

$$\mu_1\left(\frac{\partial u_1}{\partial y}\right)_{\text{interface}} = \mu_2\left(\frac{\partial u_2}{\partial y}\right)_{\text{interface}} \tag{7-1}$$

so the gradient of the velocity profile must change sharply at the interface in the ratio μ_2/μ_1. This little development is more than of just casual interest. We shall see later that it contains the germ of one of the most important ideas in wall-bounded turbulent boundary layers.

There is another important difference between laminar and turbulent boundary layer profiles on a flat plate. For laminar flow, all flat-plate profiles are the same on a plot of U/U_e versus y/δ, no matter what the fluid, the Reynolds number, the roughness size, and so on. For turbulent flows, one finds

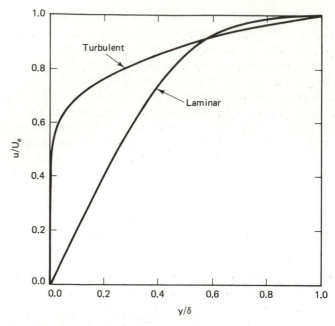

Figure 7-1 Comparison of the shapes of typical laminar and turbulent boundary layer velocity profiles.

the situation shown in Fig. 7-2. Changes in the Reynolds number and/or the roughness change C_f and that influences the profile shape in terms of U/U_e versus y/δ.

There is a consequence of the behavior shown in Fig. 7-2 that is directly important in analysis. For laminar flow the ratio $\delta^*/\theta \equiv H$, called the *shape parameter*, may be taken as describing the profile shapes that occur under different flow situations. For example, a laminar flat plate profile always has $H = 2.6$. This behavior was used in Sec. 6-2 and also in the integral analyses of Chapter 2. For turbulent flows this concept is simply not valid, although it has often been used. Referring to Fig. 7-2, it can be seen that H for a turbulent flat-plate boundary layer depends on C_f. Thus, H cannot be used to unequivocally indicate the shape of the profiles to be found for turbulent boundary layer flows for a given body (implying a specific dP/dx).

It is very important to develop the most powerful and comprehensive correlating variables possible, so that the available empirical information can be presented and studied in as compact a form as possible. The problem is complex, and there is a lot of data. To understand as much as possible about the problem, it is obviously helpful to consider the data in the simplest form involving the fewest groupings of variables and parameters. Clearly, U/U_e and y/δ are not sufficient to collapse all the data, even for simple turbulent flat-plate flow, and we will have to look deeper. Many workers contributed to the

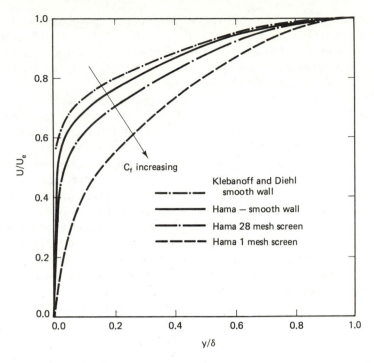

Figure 7-2 Turbulent boundary layer velocity profiles on smooth and rough flat plates. (From Clauser, 1956.)

final results that will be developed below, but the presentation by Clauser (1956) is the clearest and most comprehensive, and the general outline of his approach is followed here.

The curves in Fig. 7-2 cannot be collapsed into a single curve by multiplying by any single scaling factor. However, if the *velocity defect* defined as $(1 - U/U_e)$ is considered, a factor proportional to $1/\sqrt{C_f}$ will correlate all the curves. The velocity defect is the local decrease in velocity below the boundary layer edge velocity. Since the profiles are expressed in terms of velocities, it is convenient to introduce the derived velocity called the *friction velocity* u_* as a normalizing quantity. The wall shear divided by the density has the units of velocity squared, so one can write

$$u_* \equiv \sqrt{\frac{\tau_w}{\rho}} \qquad (7\text{-}2)$$

leading to

$$\frac{u_*}{U_e} = \sqrt{\frac{\tau_w}{\rho U_e^2}} = \sqrt{\frac{C_f}{2}} \qquad (7\text{-}2a)$$

Thus, the proper choice for the ordinate of turbulent, flat-plate boundary layer profiles can be taken as $(U/U_e - 1)/\sqrt{C_f/2} = (U - U_e)/u_*$. It suffices to retain y/δ as the other axis. The success of this choice of coordinates is shown in Fig. 7-3. This type of plot is commonly called a *defect law plot*, and the statement $(U - U_e)/u_* = f(y/\delta)$ is called the *defect law*.

Unfortunately, the great success at correlation indicated in Fig. 7-3 does not extend down to very small values of y/δ in the so-called near-wall region. As was seen earlier, the velocity changes rapidly in this region down to zero at the wall, and the variation from case to case is obscured on a plot such as Fig. 7-3. Physical reasoning can tell us that this should indeed be the case. In the outer part of the boundary layer, we saw that it was the local value of the mean velocity relative to the edge velocity $(U - U_e)$ that was important. The value of the local mean velocity relative to the wall value of zero U did not enter directly. In the innermost part of the layer, surely that measure will become important. It seems plausible to retain the friction velocity as a scaling quantity for velocity, especially in the wall region, so the trial grouping $(U/u_* \equiv u^+)$ can be selected. For the outer part of the boundary layer, location in the layer y was reckoned compared to δ as y/δ. In the inner, wall-dominated region, it can be reasoned that the value of δ will be less important. It remains

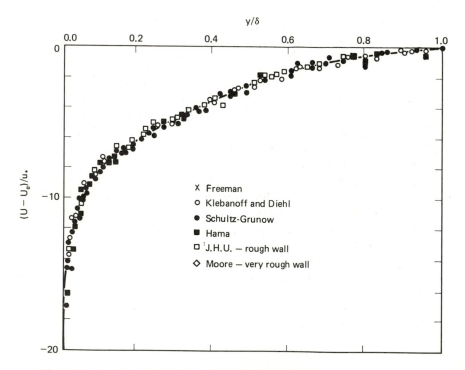

Figure 7-3 Defect law plot of turbulent velocity profiles in the outer region of the boundary layer for flow over smooth and rough flat plates. (From Clauser, 1956.)

then to nondimensionalize the length y in some other way more appropriate to that region. A little thought leads to a Reynolds number of the form

$$y^+ \equiv \frac{y u_*}{\nu} \tag{7-3}$$

The success of these variables for correlating profiles in the inner region is demonstrated in Fig. 7-4. Note that only smooth solid wall data are included on this plot. The influence of roughness will be discussed shortly. Also, the rationale for the use of the logarithmic plot and the origin of the labeled curves will be presented below. A plot such as in Fig. 7-4 is often called a *wall law plot*, and the statement $U/u_* = g(y u_*/\nu)$ is called the *law of the wall*.

At this point, it may be of interest to the reader to note that since $u_* \equiv \sqrt{\tau_w/\rho}$ plays such a key role in understanding velocity profiles, the accurate determination of the wall shear τ_w is essential. This whole matter really only became codified with the advent of successful skin friction balances in the early 1950s (see Dhawan, 1953). A device of this type has a small portion of the surface of interest that is separated from the rest of the surface (a narrow gap surrounds the measurement surface) and connected to a sensitive force-measuring system. Thus, the frictional force on a known area of the surface is measured directly. Earlier workers attempted to use the slope of the velocity profile at the wall to evaluate τ_w, but that cannot lead to accurate results because of the shape of the profile for small y/δ (see Fig. 7-1).

Let us now return to Fig. 7-4 and study the wall region further. Near a smooth, solid surface, the velocity fluctuations must be strongly damped, since right at the wall, $u' = v' = w' = 0$. Thus, there should be some small layer of laminar flow on the wall within the turbulent boundary layer. This is termed the *laminar sublayer*. This can be argued another way; if $y u_*/\nu$ is a suitable Reynolds number for describing the flow in this region, there should be some lower critical value, below which laminar, not turbulent flow prevails. This implies a lower critical value of y for a given flow situation. If such a laminar sublayer exists, it will surely be very thin. One can therefore suppose that $\tau = \tau_w$ in this region, since $\partial \tau/\partial y = 0$ at the wall where $U = V = 0$ for a flat plate [see Eq. (3-23)]. This leads to

$$\mu \frac{\partial U}{\partial y} = \tau_w \neq F(y) \tag{7-4}$$

which can be integrated to give

$$U = \frac{y \tau_w}{\mu} \tag{7-5}$$

Recast in terms of our wall law variables (u^+, y^+), Eq. (7-5) becomes simply

$$\frac{U}{u_*} = \frac{y u_*}{\nu} \tag{7-5a}$$

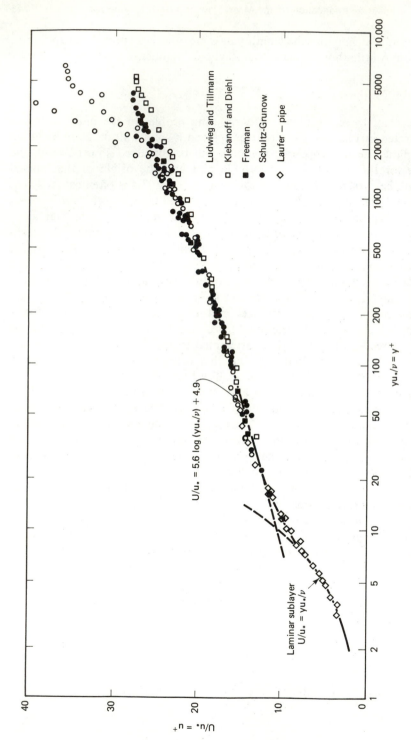

Figure 7-4 Universal wall law plot for turbulent boundary layers on smooth, solid surfaces. (From Clauser, 1956.)

146

or

$$u^+ = y^+ \tag{7-5b}$$

One can see on Fig. 7-4 that the data follow this relation out to about $y^+ \approx 7$. This confirms the existence of the laminar sublayer and shows that it is, indeed, very thin (less than 1% of δ).

Upon examining the inner and outer region plots in Figs. 7-3 and 7-4, it is found that there is a substantial region of overlap where both forms apparently apply. In such an *overlap region*, we may say that

$$\frac{U - U_e}{u_*} = f\left(\frac{y}{\delta}\right) \tag{7-6}$$

and

$$\frac{U}{u_*} = g\left(\frac{yu_*}{v}\right) \tag{7-7}$$

Equations (7-6) and (7-7) can be rewritten as

$$\frac{U}{u_*} = f\left(\frac{y}{\delta}\right) + \frac{U_e}{u_*} \tag{7-6a}$$

and

$$\frac{U}{u_*} = g\left[\left(\frac{y}{\delta}\right)\left(\frac{\delta u_*}{v}\right)\right] \tag{7-7a}$$

In the overlap region, the two must be equal for all y. For a given boundary layer profile, U_e/u_* and $\delta u_*/v$ are constants, (i.e., they do not vary with y). Thus, a constant multiplicative factor inside the function g must have the same effect as another constant additive factor outside the function f. The logarithm is the only function that displays such a property. Thus, we may propose for the *overlap region only*

$$\frac{U - U_e}{u_*} = A \log\left(\frac{y}{\delta}\right) + B \tag{7-6b}$$

and

$$\frac{U}{u_*} = A \log\left(\frac{yu_*}{v}\right) + C \tag{7-7b}$$

The *wall law plot* was already plotted on a log scale in Fig. 7-4, and the *defect law plot* from Fig. 7-3 can be replotted as such as in Fig. 7-5. Examining the data so plotted, Clauser (1956) proposed that $A = 5.6$, $B = -2.5$, and $C = 4.9$. Different lines can be drawn through the type of data shown, leading to slightly different values of the constants. Other workers have done so, but we will use Clauser's values throughout this book. Also, some workers write these

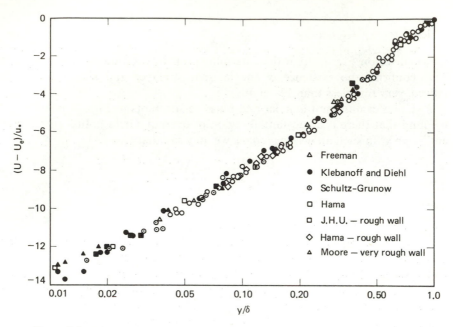

Figure 7-5 Logarithmic plot of the defect law for flat-plate turbulent boundary layers. (From Clauser, 1956.)

equations with a natural logarithm and use a constant κ, which is related to A as $\kappa = \ln (10)/A$. For $A = 5.6$, $\kappa = 0.41$.

Coles (1956) introduced the notion of a *law of the wake* to describe the velocity profiles in the outer region. He used the fact that the deviation of the velocity above the logarithmic law, when normalized with the maximum value of that deviation at the outer edge of the layer, $y = \delta$, is a function of y/δ alone. He correlated that function as a *wake function* $W(y/\delta)$, which is selected to have $W(0) = 0$ and $W(1) = 2$ as

$$\frac{U/u_* - (A \log (yu_*/v) + C)}{U_e/u_* - (A \log (\delta u_*/v) + C)} \equiv \tfrac{1}{2} W\left(\frac{y}{\delta}\right) \tag{7-8}$$

Coles (1956) then proposed that

$$W\left(\frac{y}{\delta}\right) = 2 \sin^2 \left(\frac{\pi}{2} \frac{y}{\delta}\right) \tag{7-9}$$

Finally, the *law of the wake* can be written as

$$\frac{U}{u_*} = A \log \left(\frac{yu_*}{v}\right) + C + \frac{\Pi}{\kappa} W\left(\frac{y}{\delta}\right) \tag{7-10}$$

giving a composite profile valid for both the overlap layer and the outer region. Here, Π is a *wake parameter*, $\Pi = -\kappa B/2$. For $B = -2.5$, $\Pi = 0.51$ [Coles (1956) recommended 0.55].

A skin friction law can be found by eliminating U and y from Eqs. (7-6b) and (7-7b) by subtracting to give

$$\frac{U_e}{u_*} = A \log\left(\frac{\delta u_*}{\nu}\right) + C - B \tag{7-11}$$

which can be rewritten as

$$\sqrt{\frac{2}{C_f}} = A \log\left(\text{Re}_\delta \sqrt{\frac{C_f}{2}}\right) + C - B \tag{7-12}$$

Other simpler, explicit formulas have also been derived either by approximating Eq. (7-12) or fitting empirical data directly. One of the simplest is due to Blasius as confirmed by Schultz-Grunow (1940):

$$C_f = 0.0456(\text{Re}_\delta)^{-1/4} \tag{7-13}$$

This is valid up to approximately $\text{Re}_x = 10^7$. Many people use the famous Schoenherr (1932) formula:

$$\frac{1}{\sqrt{C_f}} = 4.15 \log(\text{Re}_x \, C_f) + 1.7 \tag{7-14}$$

For the total frictional resistance coefficient C_D for turbulent flow over a flat plate of length L, these same two workers proposed

$$C_D = \frac{0.427}{(\log(\text{Re}_L) - 0.407)^{2.64}} \tag{7.13a}$$

and

$$\frac{1}{\sqrt{C_D}} = 4.13 \log(\text{Re}_L \, C_D) \tag{7-14a}$$

Surface roughness has an important direct effect on the flow in the inner, wall-dominated region. It has only an indirect effect on the outer flow by increasing C_f. The precise description of surface roughness patterns and their effects on turbulent flows cannot be based on any single parameter such as the average roughness size k. Research is still being directed at this question, and a detailed discussion is not within the scope of this book. For our purposes, we will proceed as if k alone were sufficient, except for a few comments. In the wall region, one would suppose that k enters nondimensionalized as $ku_*/\nu \equiv k^+$. For a roughness size well within the laminar sublayer, the influence can be expected to be small, so one might say that surfaces with $k^+ < 5$ (approximately) are *smooth* except for unusual roughness patterns. For surfaces with $k^+ > 10$ to 12 (approximately), the laminar sublayer begins to disappear. Experiment shows that the influence of roughness on the *wall law plot* is to shift the logarithmic portion of the smooth wall curve down and to the right, corresponding to an increase in C_f. The slope of the logarithmic region, however, does not change. The factor A (and hence also κ) in the logarithmic law is not influenced by processes in the sublayer, so it will also not be influenced by modest roughness. This result can be represented as a downward shift written as $(\Delta U/u_*)$ at a fixed (yu_*/v). This shift is a function of k^+ as shown in Fig. 7-6. The *law of the wall* is then written as

$$\frac{U}{u_*} = A \log\left(\frac{yu_*}{v}\right) + C - \frac{\Delta U}{u_*} \qquad (7\text{-}15)$$

For large k^+ where the laminar sublayer disappears ($k^+ \geq 70$), the flow is said to be *fully rough* and the inner layer must be independent of viscosity. For this to happen, we can see from Eq. (7-15) that $(\Delta U/u_*)$ must have the following form, so that the viscosity cancels:

$$\frac{\Delta U}{u_*} = A \log\left(\frac{ku_*}{v}\right) + D \qquad (7\text{-}16)$$

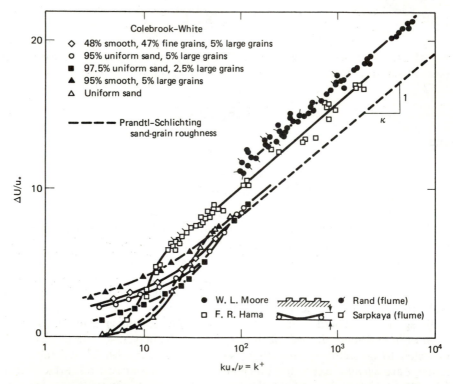

Figure 7-6 Downward shift of the logarithmic portion of the wall law as a result of roughness. (From Hama, 1954.)

If this is true, the curves in Fig. 7-6 must have the slope $A = 5.6$ for large k^+, which they do. The quantity D is the y *intercept*.

The new *law of the wall* for fully rough surfaces, Eq. (7-15) with (7-16), can be combined with the *defect law*, Eq. (7-6b) (which holds for smooth and rough surfaces) as before to determine a skin friction law. The result is

$$\sqrt{\frac{2}{C_f}} = A \log \left(\frac{\delta}{k}\right) - B + C - D \qquad (7\text{-}17)$$

The quantity D depends somewhat on the character of the roughness even for large k^+, as can be seen on Fig. 7-6. For small k^+, the situation is more complex with the type of roughness strongly influencing the flow in addition to k^+.

The state of current knowledge for turbulent boundary layer flow over permeable surfaces of the type that might be used for suction or injection applications is much poorer than for flows over solid surfaces. These cases have practical importance because of applications such as drag reduction on surfaces by maintaining laminar flow at high Reynolds numbers by suction and thermal protection by fluid injection. The poor state of current knowledge can be traced to three main sources. First, it has proven difficult to separate the influences of the small roughness attendant to most porous surfaces and the porosity itself. Second, a wide variety of porous surface types have been used in the various studies, and the behavior of the different surfaces compared to each other has not been fully clarified. Studies have been conducted with sintered metal powder surfaces, layered diffusion-bonded screening, sheets perforated with a fine array of small holes, and a layer of small spheres as examples. Third, there is the matter of the measurement of the wall shear. Most of the work in the literature has involved the use of either the slope of the velocity profile at the wall or the integral momentum equation, methods that are known to give generally inadequate accuracy even for solid, smooth walls.

Two of the more widely accepted suggestions for extending the logarithmic law to cases with suction or injection are given as follows:

Stevenson (1963) law:

$$\left(\frac{2}{v_0^+}\right)[(1 + v_0^+ u^+)^{1/2} - 1] = A \log (y^+) + C \qquad (7\text{-}18)$$

Simpson (1968) law:

$$\left(\frac{2}{v_0^+}\right)[(1 + v_0^+ u^+)^{1/2} - (1 + 11.0 v_0^+)^{1/2}] = A \log \left(\frac{y^+}{11.0}\right) \qquad (7\text{-}19)$$

where $v_0^+ \equiv v_w/u_*$.

It is interesting and troublesome that these expressions do not agree with each other, so they cannot both be correct. Each was mainly developed using each author's own data. However, one expression seems to agree best with

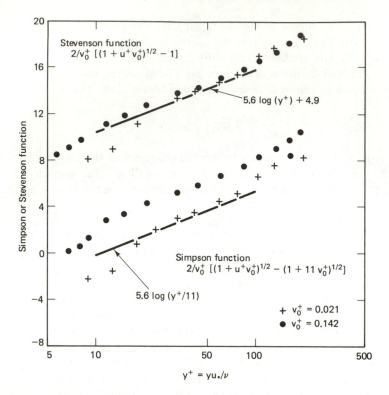

Figure 7-7 Test of the Simpson and Stevenson wall law proposals for cases with injection or suction against the data of Scott et al. (1964). (From Schetz and Favin, 1971.)

some data and the other with other data. Both suggestions have been used to attempt to correlate independent sets of data in Figs. 7-7 and 7-8. The porous surface for the experiments of Scott et al. (1964) was layered screening, whereas that used by Schetz and Nerney (1977) was sintered metal powder.

For the experimental case in Fig. 7-7, the Stevenson law appears best, but for the case in Fig. 7-8, the Simpson law looks best at least as far as collapsing all the data into a narrow band. Actually, the data of Schetz and Nerney (1977) showed a behavior more like that observed over rough solid surfaces, as may be seen in Fig. 7-9. For all cases, a shift of $\Delta U / u_* \approx 3.5$ was found in the range $0 \leq v_0^+ \leq 0.1$. It is perhaps worth noting that the wall shear in those tests was measured directly with a floating element balance.

In an attempt to sort out the influences of the many variables and parameters effecting the development of the flow in the inner region over porous surfaces, a series of controlled experiments was conducted and reported in Kong and Schetz (1981) and (1982). A total of six surfaces were studied at the same conditions in the same apparatus: (1) smooth, solid; (2) sandpaper-

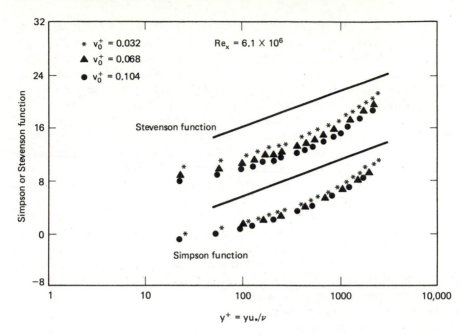

Figure 7-8   Test of the Simpson and Stevenson wall law proposals against the data of Schetz and Nerney (1977).

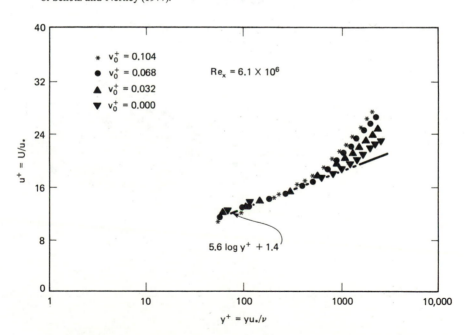

Figure 7-9   Law of the wall results for injection through a sintered metal porous surface. (From Schetz and Nerney, 1977.)

roughened, solid; (3) sintered metal powder; (4) diffusion-bonded screening, porous; (5) diffusion-bonded screening with a solid sheet bonded to the underside; and (6) a thin titanium sheet with 0.15-mm electron-beam-drilled holes on a 0.625-mm-center-to-center square pattern. The sandpaper roughness was in the same nominal range ($k^+ \sim 5$ to 7) as the sintered and screening surfaces. A typical wall law plot is shown in Fig. 7-10. The effect of roughness in this range on a solid surface can be seen in the sandpaper-roughened and solid-backed screening cases where a negligible shift in the logarithmic region $\Delta U/u_*$ was found, as would be expected for k^+ values of this magnitude. The effect of porosity can be shown by comparing the sintered metal, porous wall results to the sand-roughened, solid wall results. Although the detailed character of the roughness patterns for these two cases is different, the average k^+ is in the same range. To see the effect of porosity without any interference from different surface roughness patterns, it is clearer to compare the results between the "smooth" perforated titanium wall and the smooth, solid wall, or between the porous and solid screening walls, since the roughness patterns of these last two surfaces are exactly the same. The comparisons reveal that the effect of porosity is to shift the logarithmic region of the wall law downward by an amount $\Delta u^+ = 3$ to 4 from the solid wall results. The combined effects of small roughness and porosity can be seen by comparing the results between the sintered metal, porous wall and the smooth, solid wall or between the porous, rough screening wall and the smooth, solid wall. It is observed that the

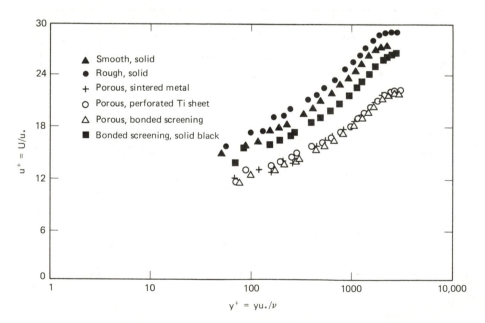

Figure 7-10 Law of the wall results for flow over various solid and porous surfaces. (From Kong and Schetz, 1982.)

downward shift of the logarithmic region of the wall law by the combined effects of small roughness and porosity is approximately the sum of the individual effects.

In general, skin friction is increased by suction and decreased by injection as shown in Fig. 7-11. However, it is important to observe that the skin friction over a porous surface is increased compared to flow over a smooth solid surface at the same conditions even for $v_w \equiv 0$. Thus, it may be necessary to inject at a significant rate before getting back down to the smooth, solid surface value. The initial increase in C_f at $v_w = 0$ was found to be approximately the sum of about a 15 to 20% increase each from roughness and porosity. The smoother perforated titanium sheet has a lower initial increase at $v_w = 0$, but the decrease as v_w increases is less than for the sintered surface. This is presumably because the injection is more uniform on the small scale for the sintered material than for the individual small holes of the perforated titanium.

7-3-2 Flow in a Pipe

Another simple, wall-bounded turbulent flow is that in a round tube or pipe. For a sufficiently long pipe (measured in L/D), the flow becomes *fully developed* such that the velocity profile does not change with distance along the pipe. Some useful results can be obtained from a simple analysis following

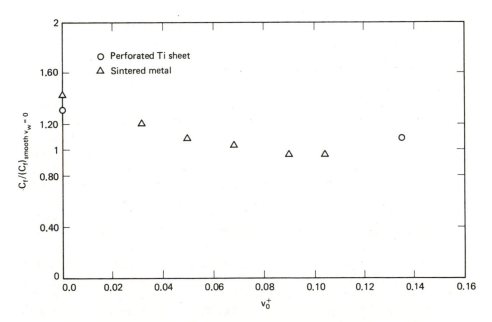

Figure 7-11 Skin friction reduction with injection through two porous surfaces. (From Schetz and Nerney, 1977, and Collier and Schetz, 1983.)

the method used in Sec. 4-2 for laminar flows. This again leads to the general result

$$\tau_w = \frac{-R}{2}\frac{dP}{dx}$$ (7-20)

Thus, the wall shear can be inferred from the pressure drop along the pipe. Before the development of the skin friction balance, the only reliable turbulent skin friction measurements were obtained in that way. As for laminar flow, the *resistance coefficient* λ [see Eq. (4-7)] is used to present data for the frictional resistance. Data for smooth pipes in the laminar, transitional, and turbulent regimes is shown in Fig. 7-12 together with some empirically based equations from Blasius (1913),

$$\lambda = 0.316(\text{Re}_D)^{-1/4}$$ (7-21)

and Prandtl (1935),

$$\frac{1}{\sqrt{\lambda}} = 2.0 \log (\text{Re}_D \sqrt{\lambda}) - 0.8$$ (7-22)

Figure 7-12 shows clearly the important fact that turbulent skin friction is generally higher than the corresponding laminar value, especially for Reynolds numbers of the order of the value for transition.

The influence of uniform roughness on the frictional resistance to flow in pipes is usually given in terms of the well-known *Moody Chart*, as shown in Fig. 7-13. Note that in this case k has been made dimensionless simply with the pipe diameter D. These results have been used to infer the corresponding information for flow over flat plates.

Turning to the question of the velocity profiles found for fully developed turbulent flow in pipes, we might expect to find something similar to the inner and outer layers found for the flat plate. Indeed, that is the case; moreover, the *law of the wall* determined for the flat-plate case has been found to apply directly for the pipe flow case also. This equivalence may seem surprising, since pipe flow has a pressure gradient ($dP/dx \neq 0$). We shall see in the next section that the implications of this happy finding are profound.

For the *defect law* in the outer portion of the layer, we cannot expect complete equivalence, since the flat-plate boundary layer is bounded by $\delta(x)$, whereas the outer (away from the wall) edge of the pipe flow layer is found at a constant distance R (measured from the wall). The transposition is, however, simple. One could almost guess that the velocity scale should be $(U - U_{max})/u_*$ with the distance scale changed from y/δ to y/R, and this is found to be the case, as shown by the data in Fig. 7-14. Again, this *defect law* is valid for both smooth and rough surfaces.

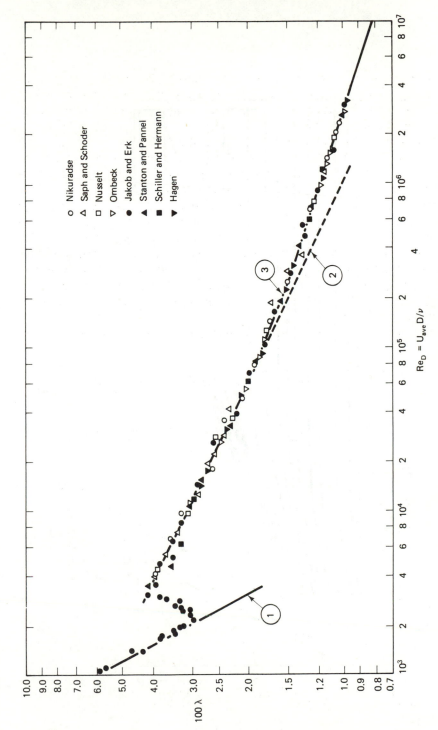

Figure 7-12 Frictional resistance in a smooth pipe. Curve 1, Eq. (4-7a); curve 2, Eq. (7-21); curve 3, Eq. (7-22). (Data collected by Schlichting, 1968.)

Legend:
- ○ Nikuradse
- △ Saph and Schoder
- □ Nusselt
- ▽ Ombeck
- ● Jakob and Erk
- ▲ Stanton and Pannel
- ■ Schiller and Hermann
- ▶ Hagen

Vertical axis: $100\,\lambda$

Horizontal axis: $Re_D = U_{ave}D/\nu$

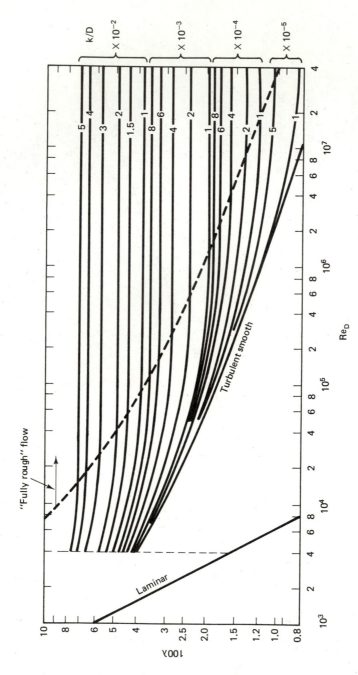

Figure 7-13 Frictional resistance in uniformly roughened pipes. (From Moody, 1944.)

158

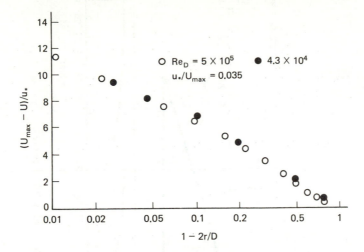

Figure 7-14   Modified defect law plot for the outer region of a pipe boundary layer. Open circles: Laufer (1954); solid circles: Nikuradse (1932).

7-3-3 Flows with Axial Pressure Gradients

We have succeeded in constructing a more-or-less tidy picture of the mean flow for the two simple cases of a flat plate and a pipe, but it is now necessary to turn to more general cases. For example, consider the flow over an airfoil. Each such case will have a different pressure gradient, dP/dx, depending on the shape of the section. Can we really expect to collapse all the profiles along the surface into a single shape as for the *defect law* for flat-plate flow? After all, that is not possible even for laminar flows with pressure gradients; only the highly restrictive *similar, wedgeflow* solutions exhibit that behavior. Measurements of boundary layer profiles along the top and bottom surfaces of an airfoil are shown in Fig. 7-15, and they confirm the reservations stated. The profile shapes change markedly along the surface, and no simple scaling will bring them into coincidence. The same behavior is observed for flows over axisymmetric bodies and in planar or axisymmetric channels of varying cross section. The simple fact is that general body or channel shapes produce general pressure gradients which lead to turbulent velocity profiles whose shapes do not fall into any one simple *family*.

The assessment developed above holds in general, and the issue will have to be confronted directly below. There is, however, one simplifying experimental result that cannot be easily detected in the type of profiles shown in Fig. 7-15. The inner flow has been found to be amazingly insensitive to pressure gradients (see Ludwieg and Tillmann, 1950) right up to those approaching sufficient strength to induce separation. A hint of this came in Sec. 7-3-2, where it was observed that the pressure gradients in pipe flows did not change the *law of the wall*, but the actual result is much more general. In essence, one may

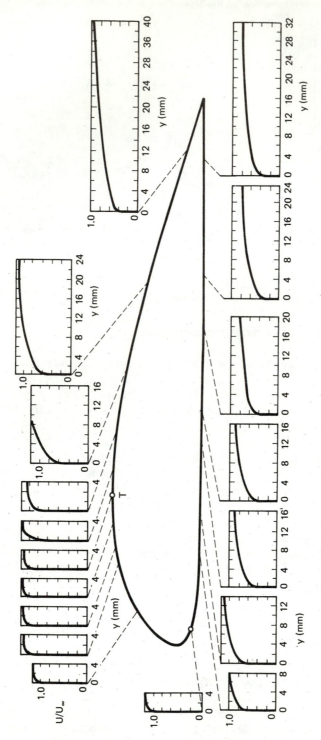

Figure 7-15 Boundary layer profiles along an airfoil: $C_L = 0.4$, $Re_L = 4.3 \times 10^6$. Pressure side turbulent, suction side turbulent beyond T. (From Stüper, 1934.)

take the law of the wall in Fig. 7-4 as holding for all pressure gradients, except in the immediate vicinity of separation. This will be seen later to have far-reaching implications.

The fact that the observable profile shape such as in Fig. 7-15 does not have a simple, universal behavior in the presence of general pressure gradients directly influences the correlation scheme introduced earlier for the outer layer of the flat-plate case, namely the *defect law*. We cannot expect all the profiles in Fig. 7-15, for example, to collapse to one profile when plotted in terms of a *defect law plot*, $(U - U_e)/u_*$ versus y/δ. However, Clauser (1954) cleverly turned this question around. He sought to find flows with nonzero pressure gradients that would produce a single profile on a *defect law plot*. Such special flows would be the turbulent equivalent of the *wedgeflow, similar* pressure gradient cases for laminar flows. Each would produce a single profile (suitably scaled) for all axial locations in the flow. For laminar flows, the scaling is u/U_e versus y/δ, and for the outer part of a turbulent boundary layer the scaling is $(U - U_e)/u_*$ versus y/δ. If the profile shape on a suitable plot is viewed as an *output* and some variable or parameter involving the pressure gradient is the *input*, we seek to find the appropriate form of the input which when held constant will result in a constant output. Clauser called this class of turbulent boundary layer flows *equilibrium pressure gradient flows*. The condition for such pressure gradients in a turbulent flow was found by reinterpreting and extending the corresponding laminar case of the wedge flows. Clauser (1956) made dP/dx dimensionless by scaling with δ and τ_w to form a new grouping $\delta/\tau_w\,(dP/dx)$, which was found to be a constant for any wedge flow in the laminar case. The length scale could as well have been δ^* or θ, since δ^*/δ and θ/δ are simply constants for any wedge flow. For turbulent flow, δ^*/δ and θ/δ depend on C_f even for a flat-plate flow, so only one choice will serve. Clauser (1954) found the proper choice to be $\delta^*/\tau_w(dP/dx)$, and he laboriously produced two such flows, as shown in Fig. 7-16. The flows, denoted as pressure gradient 1 and 2, are *equilibrium pressure gradients* (in these cases—adverse pressure gradients, $dP/dx > 0$), since they each produce a single profile on a *defect law plot*. Other examples have since been found, but this class of flows specified by constant values of $\delta^*/\tau_w(dP/dx)$ contains the only relatively simple turbulent boundary layer cases with pressure gradients known.

Coles (1956) extended the *law of the wake* to *equilibrium pressure gradients* by letting

$$\Pi = \Pi\!\left(\frac{\delta^*}{\tau_w}\frac{dP}{dx}\right) \tag{7-23}$$

White (1974) has proposed that

$$\Pi = 0.8\left[\frac{\delta^*}{\tau_w}\frac{dP}{dx} + 0.5\right]^{3/4} \tag{7-24}$$

This enables the development of a skin friction law for these cases by evalu-

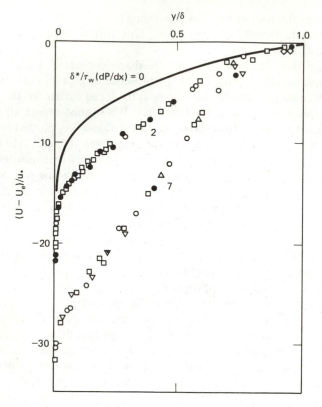

Figure 7-16   Defect law plot of velocity profiles for equilibrium pressure gradient flows. (From Clauser, 1956.)

ating the *law of the wake*, Eq. (7-10), at $y = \delta$, where $U = U_e$:

$$\sqrt{\frac{2}{C_f}} = A \log \left(\text{Re}_\delta \sqrt{\frac{C_f}{2}} \right) + C + \frac{2\Pi}{\kappa} \qquad (7\text{-}25)$$

where an equation such as Eq. (7-24) is implied.

7-4 SELECTED EMPIRICAL TURBULENCE INFORMATION

For illustrative purposes, the excellent data obtained at the National Bureau of Standards (NBS) for a turbulent boundary layer on a flat plate in the 1950s have been selected as the principal data to be presented. There were two primary reasons for this choice. First, the basic flow is simple, yet representative. Second, a complete range of measurements is available for the same flow. Another possible choice would have been the pipe flow measurements of

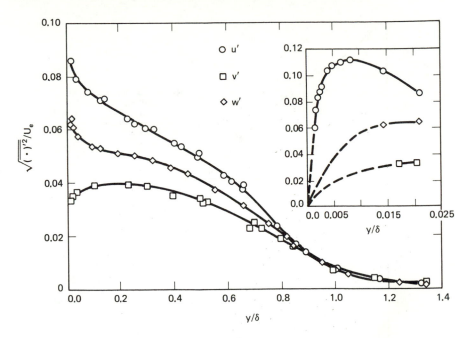

Figure 7-17 Turbulence intensity profiles for flow over a flat plate. (From Klebanoff, 1954.)

Laufer (1954). We shall refer to those here only as they compare to the flat-plate data under discussion.

Some of the NBS data from Klebanoff (1954) has already been given in Fig. 1-10. In Fig. 7-17 we show more detailed profiles in the wall region for the three turbulence intensities. Notice that the axial intensity ($\sqrt{\overline{u'^2}}/U_e$) is the largest, but that the out-of-plane intensity ($\sqrt{\overline{w'^2}}/U_e$) is larger than the transverse intensity ($\sqrt{\overline{v'^2}}/U_e$). It is also of interest to have an idea of the variations about the rms averages. Blackwelder and Kaplan (1976) have determined that the probability of finding $u' \approx \sqrt{\overline{u'^2}}$ is about 0.3 and that for finding $u' \simeq 3\sqrt{\overline{u'^2}}$ is 0.01. The profiles found by Laufer (1954) for a pipe flow resemble these closely except near the outer (away from the wall) edge of the boundary layer, which for the pipe flow is the centerline. There, in a pipe, all the intensities drop to a value of about 0.03 rather than the essentially zero value found for the flat plate as $(y/\delta) \to 1$. Also, the peak values very close to the wall are somewhat lower for the flow in a pipe.

Turbulence intensity profiles for flow over a rough flat plate from Corrsin and Kistler (1954) are plotted in Fig. 7-18. The roughness was in the form of corrugated paper, so it was two-dimensional and approximately sinusoidal. The height of the roughness was 0.2 cm, which was sufficient to put the flow into the *fully rough* regime. The turbulence intensity levels found are much

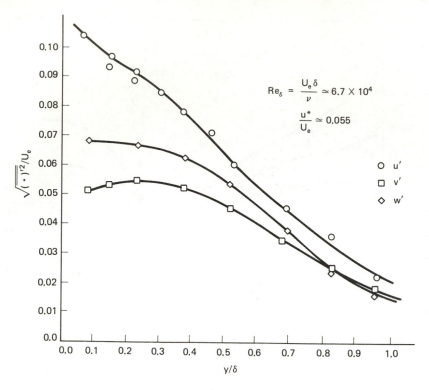

Figure 7-18 Turbulent intensity profiles for flow over a *fully rough* flat plate. (From Corrsin and Kistler, 1954.)

higher than those for the smooth wall, but the relative ordering of the three components is the same.

The outer edge of a turbulent boundary layer is characterized by an unsteady, *turbulent front*. A hot-wire anemometer placed at $y = \delta$ (recall that δ is a *mean* flow quantity) will sense alternately turbulent and then inviscid flow. This is reflected in a quantity Ω called the *intermittency* or the fraction of the time that the flow is turbulent. Some measurements are shown in Fig. 7-19. Note that the intermittency remains significantly less than unity rather far down into the boundary layer, indicating that the transverse motion of the turbulent front is on a scale which is appreciable compared to δ.

The transport of x momentum across the boundary layer by the turbulence is through the correlation $\overline{u'v'}$. When formed as $-\rho\overline{u'v'}$, this has the dimensions of a stress, and it is called the *turbulent shear stress* or the *Reynolds stress* after Osborne Reynolds. This concept will be derived more directly in a later section, but for now it is enough to observe that the primary effect of the turbulent fluctuations on the boundary layer is through the Reynolds stress. A plot of measurements of that quantity normalized with the friction velocity

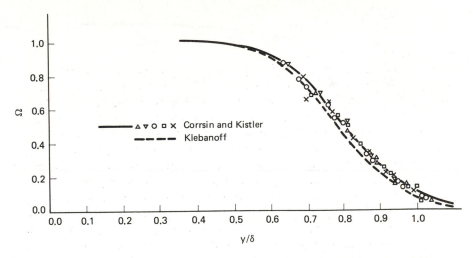

Figure 7-19 Intermittency distribution in the boundary layer. (From Klebanoff, 1954, and Corrsin and Kistler, 1954.)

across the layer is given in Fig. 7-20. It is highest near the wall, but then drops off to zero in the thin laminar sublayer very near the wall. This is shown in Fig. 7-21. Also shown on Fig. 7-20 is the variation of the *turbulent kinetic energy* (often written simply as TKE), which is the kinetic energy (per unit

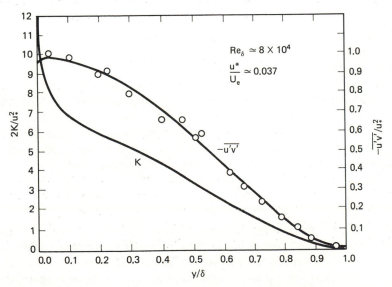

Figure 7-20 Profiles of Reynolds stress and turbulent kinetic energy in a turbulent boundary layer on a flat plate. (From Klebanoff, 1954.)

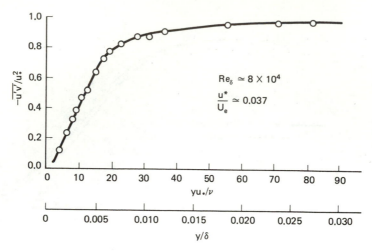

Figure 7-21 Detailed distribution of Reynolds stress in the wall region of a boundary layer. (From Schubauer, 1954.)

mass) of the three *fluctuating* velocity components.

$$K' \equiv \frac{u'^2 + v'^2 + w'^2}{2}$$

$$K \equiv \frac{\overline{u'^2 + v'^2 + w'^2}}{2} \qquad (7\text{-}26)$$

If any one quantity is to be used to describe how *turbulent* the flow is, this is the most logical choice. The profiles of the *Reynolds stress* and TKE in a pipe are quite similar to those for the flat-plate boundary layer.

An attempt has been made to emphasize for the reader the fact that the turbulent fluctuations occur over a wide range of time scales. To clarify that, we present in Fig. 7-22 some measurements of the *spectra* of the axial turbulent fluctuations as a function of the *wave number* k_1 [wave number $\equiv (2\pi) \times$ (frequency)/(mean velocity)]. The quantity $E_1(k_1)$ is the amount of turbulent energy associated with the axial component in the band k_1 to $k_1 + dk_1$. Thus,

$$\overline{u'^2} = \int_0^\infty E_1(k_1)\, dk_1 \qquad (7\text{-}27)$$

Observe first that there is appreciable motion occurring over about five decades in wave number. A second point is that the energy in the larger eddies (lower wave numbers) decreases as the wall is approached (look at the data for $y/\delta = 0.0011$), since the rigid wall inhibits larger scale motions. Again, the pipe flow results are similar to those shown here for the flat-plate case.

In any turbulent shear flow, there is *production, convection, diffusion,* and *dissipation* of turbulent energy. The *dissipation* occurs primarily due to the

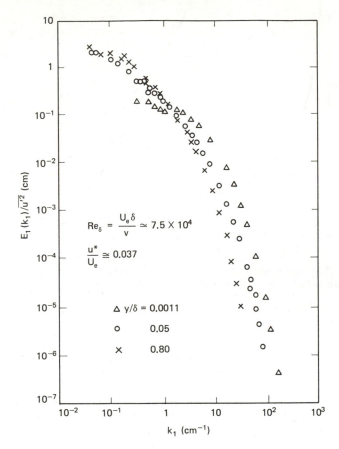

Figure 7-22 Spectra of the axial turbulence intensity in the boundary layer on a flat plate. (From Klebanoff, 1954.)

action of laminar viscosity on the smaller eddies. It is difficult to measure this quantity, but some results from the NBS studies are shown in Fig. 7-23. It is a common practice to adjust such results for the effect of intermittency to see the distribution in the fully turbulent regions of the flow.

The distributions of all the above-named processes across the layer can be plotted in the form of a turbulent kinetic energy balance as in Fig. 7-24. Except near the outer edge, the main contributions to the overall balance are from *production* and *dissipation*, which nearly counterbalance each other. It is again noted that the pipe flow results are quite similar.

For flow over permeable surfaces, the state of current knowledge of turbulent quantities is again weaker than for solid surfaces as was the case for the mean flow. Some data for axial and normal turbulence intensities and Reynolds stress are shown in Figs. 7-25, 7-26, and 7-27 for the six surfaces (three solid and three permeable) discussed earlier for the mean flow. The rough solid surfaces

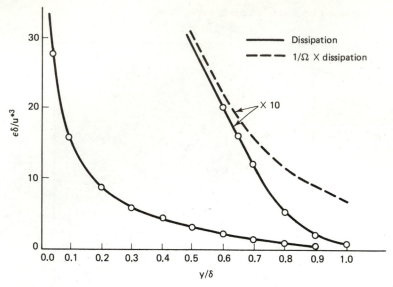

Figure 7-23 Distribution of the dissipation of turbulent kinetic energy across a flat-plate boundary layer. (From Klebanoff, 1954.)

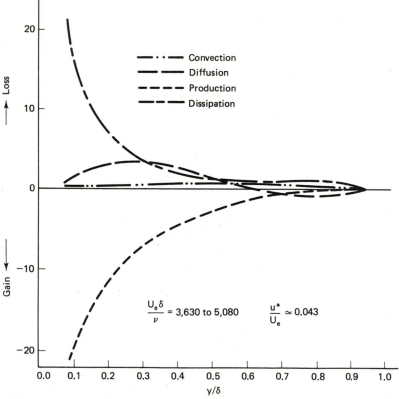

$$\frac{U_e \delta}{\nu} = 3,630 \text{ to } 5,080 \qquad \frac{u^*}{U_e} \simeq 0.043$$

Figure 7-24 Turbulent energy balance in the boundary layer on a smooth flat plate. (From Klebanoff, 1954.)

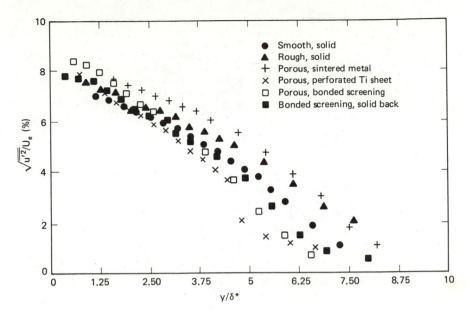

Figure 7-25 Axial turbulence intensity profiles over several solid and porous surfaces. (From Kong and Schetz, 1982.)

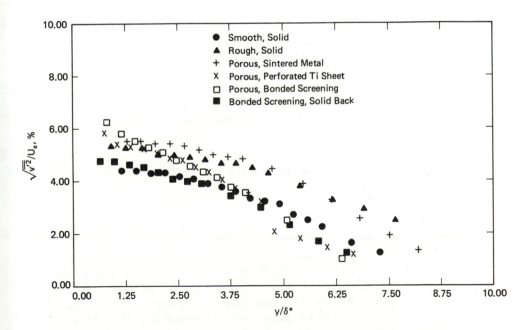

Figure 7-26 Normal turbulence intensity profiles over several solid and porous surfaces. (From Kong and Schetz, 1982.)

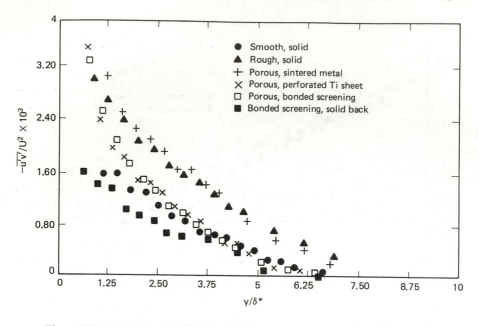

Figure 7-27 Reynolds stress profiles over several solid and porous surfaces. (From Kong and Schetz, 1982.)

show clear increases above the smooth, solid wall values for all three quantities. The effects of porosity *per se* can be seen by comparing the solid to the porous screening or the smooth solid to the smooth perforated titanium sheet. There appears to be an identifiable effect of the porosity, and it is largest for the normal turbulence intensity and the Reynolds stress. That makes sense on physical grounds, since the normal turbulence component will not be as strongly damped by a porous surface as by a solid surface.

7-5 THE CENTRAL PROBLEM OF THE ANALYSIS OF TURBULENT FLOWS

If it is assumed that the unsteady, laminar equations contain the essence of turbulent motions, then three-dimensional versions of Eq. (3-27a) and Eq. (3-6a) will serve for a constant-density, constant-property (laminar properties) case. This assumption is neither as controversial nor as helpful as it may seem. The derivation of these equations did not in any way preclude the existence of a randomly fluctuating motion. However, this assumption is not particularly helpful, since it really brings us no simplification. One is still confronted with a three-dimensional, randomly unsteady motion with very large ranges in physical scales and frequencies.

Earlier, we said that the analyst is usually interested only in the time-averaged, mean motion. How does that general goal fit into a formulation based on an unsteady system of equations? The answer is simple (in principle). Take the time average. Using our division of the motion into mean and fluctuating parts, as, for example,

$$u(x, y, z, t) = U(x, y) + u'(x, y, z, t) \tag{1-20a}$$

for a planar, steady (*in the mean*) case, we can substitute into Eqs. (3-27a) and (3-6a) and then take the time average term by term. We will work with the two-dimensional equations, since we anticipate a two-dimensional mean flow. A few simple rules must also be followed. If f and g are any two, fluctuating variables and F and G are their mean values and an overbar, $(\overline{\cdot})$, denotes the operator of taking the time mean, then

$$
\begin{aligned}
\overline{f'} &\equiv 0, & \overline{F} &\equiv F \\
\overline{f + g} &= F + G, & \overline{F \cdot g} &= F \cdot G \\
\overline{\frac{\partial f}{\partial s}} &= \frac{\partial F}{\partial s}, & \overline{\int f \, ds} &= \int F \, ds
\end{aligned}
\tag{7-28}
$$

Substituting into Eq. (3-6a), there results

$$
\begin{aligned}
0 &= \overline{\frac{\partial u}{\partial x} + \frac{\partial v}{\partial y}} = \overline{\frac{\partial (U + u')}{\partial x} + \frac{\partial (V + v')}{\partial y}} \\
&= \left(\frac{\partial U}{\partial x} + \frac{\partial V}{\partial y} \right) + \overline{\left(\frac{\partial u'}{\partial x} + \frac{\partial v'}{\partial y} \right)}
\end{aligned}
\tag{7-29}
$$

Taking the time mean

$$\frac{\partial U}{\partial x} + \frac{\partial V}{\partial y} = 0 \tag{7-30}$$

which says that the mean flow continuity equation has the same form as for a laminar flow. Note, however, that Eq. (7-30) also implies that

$$\frac{\partial u'}{\partial x} + \frac{\partial v'}{\partial y} = 0 \tag{7-29a}$$

from Eq. (7-29). Thus, the instantaneous fluctuations satisfy a continuity equation of the same form.

The nonlinear terms in Eq. (3-27a) must be treated somewhat more carefully. It is helpful to note that the convective derivative (the nonlinear terms) can be rewritten as

$$u \frac{\partial u}{\partial x} + v \frac{\partial u}{\partial y} = \frac{\partial(u^2)}{\partial x} + \frac{\partial(uv)}{\partial y}$$

$$= u \frac{\partial u}{\partial x} + v \frac{\partial u}{\partial y}$$

$$+ u \left(\frac{\partial u}{\partial x} \cancel{+ \frac{\partial v}{\partial y}} \right)^{0} \tag{7-31}$$

by virtue of the continuity equation [Eq. (7-29)]. Now substitute in terms of mean and fluctuating quantities

$$\frac{\partial(u^2)}{\partial x} + \frac{\partial(uv)}{\partial y} = \frac{\partial(U + u')^2}{\partial x} + \frac{\partial((V + v')(U + u'))}{\partial y}$$

$$= \frac{\partial(U^2 + 2Uu' + u'^2)}{\partial x} + \frac{\partial(UV + Vu' + Uv' + u'v')}{\partial y} \tag{7-32}$$

Taking the time average results in

$$\frac{\partial(U^2)}{\partial x} + \frac{\partial(UV)}{\partial y} + \frac{\partial(\overline{u'^2})}{\partial x} + \frac{\partial(\overline{u'v'})}{\partial y} \tag{7-32a}$$

Using the time-averaged continuity equation, this can be rewritten back in more familiar form as

$$U \frac{\partial U}{\partial x} + V \frac{\partial U}{\partial y} + \frac{\partial(\overline{u'^2})}{\partial x} + \frac{\partial(\overline{u'v'})}{\partial y} \tag{7-32b}$$

The other terms in the momentum equation can be treated easily:

$$\frac{\overline{\partial u}}{\partial t} = \frac{\overline{\partial(U + u')}}{\partial t} = \frac{\partial U}{\partial t}$$

$$\frac{\overline{dp}}{dx} = \frac{\overline{d(P + p')}}{dx} = \frac{dP}{dx}$$

$$\mu \frac{\overline{\partial^2 u}}{\partial y^2} = \mu \frac{\overline{\partial^2(U + u')}}{\partial y^2} = \mu \frac{\partial^2 U}{\partial y^2} \tag{7-33}$$

Thus, the time-averaged momentum equation for turbulent flow becomes

$$\frac{\partial U}{\partial t} + U \frac{\partial U}{\partial x} + V \frac{\partial U}{\partial y} + \frac{\partial(\overline{u'^2})}{\partial x} = -\frac{1}{\rho} \frac{dP}{dx} + v \frac{\partial^2 U}{\partial y^2} - \frac{\partial(\overline{u'v'})}{\partial y} \tag{7-34}$$

If the flow is steady (*in the mean*), the first term drops out. It is a common assumption to take $\partial(\overline{u'^2})/\partial x \ll \partial(\overline{u'v'})/\partial y$, and this is confirmed by experiment in many flows. Finally, the simplest system of equations that can possibly represent the mean motion of a turbulent flow may be written

$$\frac{\partial U}{\partial x} + \frac{\partial V}{\partial y} = 0 \qquad (7\text{-}30)$$

$$U \frac{\partial U}{\partial x} + V \frac{\partial U}{\partial y} = -\frac{1}{\rho}\frac{dP}{dx} + v\frac{\partial^2 U}{\partial y^2} - \frac{\partial (\overline{u'v'})}{\partial y} \qquad (7\text{-}34a)$$

Far from any rigid surface, the laminar shear term may be dropped. Certainly, the term involving $(\overline{u'v'})$ cannot be dropped, or the system would collapse completely back to that for a laminar flow.

As for any boundary layer problem, in the problem specified by Eqs. (7-30) and (7-34a), the pressure field $P(x)$ is taken as *imposed* on the shear flow by an external, inviscid flow, so that we would expect to solve these two equations for the two unknowns (U, V). However, we notice an additional unknown term as the last on the right-hand side of Eq. (7-34a). This is a momentum transfer term that plays the same role as simple Newtonian shear in a laminar flow, so that the grouping $(-\rho\overline{u'v'}) \equiv \tau_T$ is termed the *turbulent shear stress* or, as we said earlier, *Reynolds stress*. Note that this turbulent *shear* term did not come from the laminar viscous terms but from the inviscid convective derivative terms. Working with the framework of Eqs. (7-30) and (7-34a), it is necessary to relate this term to the other independent or dependent variables in order to complete the problem from a mathematical standpoint. Since these other variables are all mean quantities, such a relationship or model is termed a *mean flow model*. The development of models of that type will be the subject of Sec. 7-8.

The reader must understand that there is no easy way out of the quandary stated in the paragraph above. The matter will be pursued to more complex formulations at the end of this chapter, but there will always be more unknowns than equations. The extra unknown(s) will always be turbulent quantities such as $\overline{u'v'}$, and it will always be necessary to insert extra, generally semiempirical relations in order to *close* the system mathematically. This is called the *closure problem* for turbulent flows. It might be easier to grasp how we arrived at this perplexing state if one reflects on the simple fact that really nothing substantive has been done either by asserting definitions of mean flow quantities as by equations such as Eq. (1-20) or by manipulating the equations of motion or by taking the time average as has been done here. To this point in developing the mathematical formulation, nothing has been said about the real *nature* of turbulent flows. The equations are still waiting for us to do so, before the formulation can be complete.

Prior to proceeding with the task of developing models (i.e., *modeling* turbulent shear) it is useful to develop a physical description of the process of momentum transfer by the fluctuations. Consider the sketch in Fig. 7-28. If a fluid particle from a height $y_0 + \Delta y$ moves in an eddy down to y_0, it will tend to retain its mean velocity $U(y_0 + \Delta y)$, which will appear as a positive fluctuation on the mean velocity at y_0 [i.e., $U(y_0)$]. The instantaneous mass flow

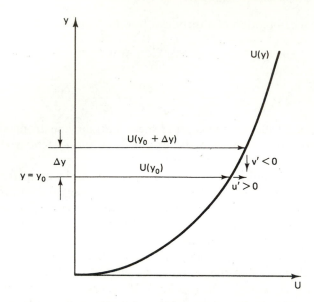

Figure 7-28   Schematic illustration of the evolution of the Reynolds stress.

through the plane $y = y_0$ will be $-\rho v'$, and its axial velocity will be $U(y_0) + u'$, so it will transport an increment of axial momentum across the plane $y = y_0$ in the amount $(-\rho u'v')$. If the particle came from below, extending the logic above leads to the same result. Averaging over time, we find that the *net* transport of axial momentum in the transverse direction by the velocity fluctuations is $-\rho \overline{u'v'}$.

7-6 MEAN FLOW TURBULENT TRANSPORT FORMULATIONS

In general, to develop a mean flow turbulent analysis, it is first necessary to *model* the turbulent shear at each point across the layer. Recalling the discussion of Sec. 7-5, this means that a relation

$$\tau_T \equiv -\rho \overline{u'v'} = f(x, y; \ U, V, P \quad \text{or} \quad U_e, \mu, \rho) \qquad (7\text{-}35)$$

must be sought so that the system posed by Eqs. (7-30) and (7-34a) can be closed. Clearly, this must be based on empirical information, so we are led back to the material in Sec. 7-3 for mean flow data. Note that no turbulent quantities appear on the right-hand side of Eq. (7-35). Perhaps it will be helpful at this point to state clearly the goal of this part of the *modeling* efforts using the inner region as an example. We wish to find a relationship as schematically written in Eq. (7-35) that will, when used in the equations of motion, Eqs. (7-30) and (7-34a), faithfully reproduce the solid curve through the data in Fig. 7-4. An analysis of this type is often called *semiempirical.* An *empirical* approach would employ curves simply fitted to the data without recourse to any equations based on first principles that govern the problem.

The goal stated above has been accomplished within two related formulations. The first is the *eddy viscosity formulation* introduced by Boussinesq (1877), which takes the form

$$\tau_T = -\rho\overline{u'v'} = \mu_T \frac{\partial U}{\partial y} \qquad (7\text{-}36)$$

by analogy with laminar flow [see Eq. (1-2)]. Here, however, μ_T, the *eddy viscosity*, must be expected to be dependent upon the state of the *flow* not just the state of the *fluid.* Thus, it is not a thermophysical property of the fluid alone as is μ, the laminar viscosity. The second formulation was suggested by Prandtl (1925) as

$$\tau_T = -\rho\overline{u'v'} = \rho\ell_m^2 \left|\frac{\partial U}{\partial y}\right| \frac{\partial U}{\partial y} \qquad (7\text{-}37)$$

where ℓ_m is a *mixing length,* so this is called the *mixing-length formulation.* Prandtl's development of the mixing length concept follows closely on the physical interpretation of $-\rho\overline{u'v'}$ at the end of Sec. 7-5 with Fig. 7-28. The *mixing length* ℓ_m is crudely similar to the *mean free path* between molecules λ^*, in that it is taken as some *effective interaction distance,* except that it is between eddies rather than molecules. If in Fig. 7-28, we take ℓ_m rather than Δy as the distance which the fluid particle moves, it will induce a velocity perturbation

$$u'(y_0) = U(y_0 + \ell_m) - U(y_0) \approx \ell_m \frac{\partial U}{\partial y} \qquad (7\text{-}38)$$

with $v'(y_0) < 0$. If the fluid particle comes from below, it will induce a velocity perturbation

$$u'(y_0) = U(y_0 - \ell_m) - U(y_0) \approx -\ell_m \frac{\partial U}{\partial y} \qquad (7\text{-}39)$$

with $v' > 0$. Thus, by continuity, one can say $v' \sim -u'$, and this with Eqs. (7-38) and (7-39), leads directly to Eq. (7-37). We write

$$\left|\frac{\partial U}{\partial y}\right| \frac{\partial U}{\partial y}$$

rather than $(\partial U/\partial y)^2$ to ensure the proper sign for τ_T in relation to that of $\partial U/\partial y$. An additional relationship for ℓ_m as a function of the independent variables or parameters or mean dependent variables is still required. It will also be a function of the state of the *flow* and not the *fluid,* just as was the eddy viscosity μ_T.

At this stage, it is appropriate to make some observations about the

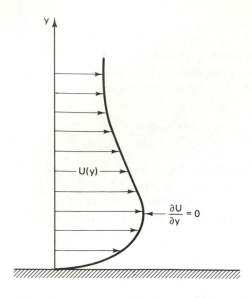

Figure 7-29 Illustration of a turbulent boundary layer velocity profile with a velocity maximum (i.e., $\partial U/\partial y = 0$).

formulations in Eqs. (7-36) and (7-37). First, both the *eddy viscosity* and *mixing-length formulations* are what are called *gradient transport formulations* since they employ a gradient (here $\partial U/\partial y$) of some relevant dependent variable to represent a transport process. Thus, our representations of laminar transport of momentum, $\tau = \mu(\partial u/\partial y)$, and thermal energy, $q = -k(\partial T/\partial y)$, are also *gradient transport formulations*. There is, however, a special weakness of such formulations for turbulent shear flows that is most easily seen for the case of a profile such as that sketched in Fig. 7-29. Profiles of this type occur, for example, with high-velocity injection through an upstream slot in the wall into a turbulent boundary layer. Since the mean velocity profile $U(y)$ has a maximum, it must have a point where $\partial U/\partial y = 0$. Both formulations imply $\tau_T = -\rho\overline{u'v'} \equiv 0$ at such a point, and that is not borne out by experiment. There is no simple way to repair this basic weakness of these formulations, but one can take some small comfort in the fact that most boundary layer flows do not involve profiles with a maximum or minimum. This does not apply for a symmetrical profile where $\partial U/\partial y = 0$ on the plane of symmetry and $-\rho\overline{u'v'} = 0$ there also. The second general observation to be made about these two formulations is that they are essentially equivalent. Looking at Eqs. (7-36) and (7-37), one can obviously write

$$\mu_T = \rho\ell_m^2 \left|\frac{\partial U}{\partial y}\right| \qquad (7\text{-}40)$$

Thus, if a relationship for μ_T, for example, is known, a corresponding relationship for ℓ_m can be found from Eq. (7-40), or vice versa.

Specific models for the eddy viscosity and the mixing length will be developed in Sec. 7-8. Before that, however, we shall look at integral methods for turbulent boundary layers.

7-7 MEAN FLOW INTEGRAL METHODS

There are methods for turbulent flow analysis that follow along the lines of the integral methods for laminar flows discussed in Chapter 2. They bear roughly the same relation to turbulent analyses based on differential formulations as is the case for laminar flows. Such methods are generally simple to apply, and they yield results that are sometimes adequate for engineering work. For purposes of illustration, we will describe two of these methods here; an older one that is suitable only for very simple problems and a newer one due to Moses (1969) that can be used for more general flows. In each case, the reader should be alert to see how the necessary modeling of the turbulent shear enters.

The earliest application of an integral method to a turbulent case was by Prandtl (1927) for the flat-plate problem. The momentum integral equation, Eq. (2-14), originally derived in Chapter 2 for steady, laminar flows, holds here for the mean flow of a steady (*in the mean*) turbulent flow if the dependent variables are all interpreted as mean values. Thus, for a solid-flat plate we have

$$\frac{d\theta}{dx} = \frac{C_f}{2} \qquad\qquad (2\text{-}14a)$$

if in the definition of θ, Eq. (2-12), we understand that the profile is described by $U(y)$, the mean velocity.

As for the laminar problem, an assumed profile shape is needed. The mean velocity profile for a turbulent flow over a flat plate can be crudely represented by

$$\frac{U}{U_e} = \left(\frac{y}{\delta}\right)^{1/n} \qquad\qquad (7\text{-}41)$$

with $n \approx 7$ for $\text{Re}_x \approx 10^6$ to 10^7. It is known, however, that this cannot be true in general because of the data and discussion in Sec. 7-3-1 (see also Fig. 7-2). Nonetheless, we proceed with Eq. (7-41), which gives $\delta^*/\delta = \frac{1}{8}$ and $\theta/\delta = \frac{7}{72}$ for $n = 7$. It might be expected that we would next use the profile to find the wall shear as for the laminar case, but this involves $(\partial U/\partial y)_w$, which from Eq. (7-41) is

$$\left.\frac{\partial U}{\partial y}\right|_w = \frac{1}{7}\frac{U_e}{\delta^{1/7}y^{6/7}}\bigg|_{y=0} = \infty \qquad\qquad (7\text{-}42)$$

However, one should have anticipated a need to insert some extra information about the shear in a turbulent flow anyway. In this method, we need only the shear at the wall, C_f in terms of δ [i.e., $C_f(\text{Re}_\delta)$], since δ is to emerge as the primary, dependent variable. For the laminar analysis, δ played the same role even though the wall shear was handled differently. For the required $C_f(\text{Re}_\delta)$, Prandlt (1927) used Eq. (7-13). Using that and $\theta/\delta = \frac{7}{72}$ in Eq. (2-14a), we obtain

$$\tfrac{7}{72} \rho U_e^2 \frac{d\delta}{dx} = \frac{0.0456}{2} \rho U_e^2 \left(\frac{\nu}{U_e \delta}\right)^{1/4} \qquad (7\text{-}43)$$

This can be rearranged and solved to give

$$\frac{\delta(x)}{x} = \frac{0.375}{\mathrm{Re}_x^{1/5}} \qquad (7\text{-}44)$$

where $\delta(0) = 0$ was used. This result indicates that a turbulent boundary layer grows much faster $(\delta \sim x^{4/5})$ than a laminar one, which has $\delta/x \sim \mathrm{Re}_x^{-1/2}$ $(\delta \sim x^{1/2})$.

Unfortunately, there is no more that can be done with this method. For example, if one wishes to solve problems with pressure gradients, it would be necessary to have $C_f(\mathrm{Re}_\delta, dP/dx)$ for general forms of dP/dx. If that much information were actually available, one would not really need this type of analysis.

An integral method which has greater capability than that above and which is still simple to apply has been developed by Moses (1969). This method uses a generalized form of the momentum integral equation, Eq. (2-14), where U is understood for u, written as

$$\frac{U(y_1/\delta)}{U_e} \frac{d}{dx} \left[\mathrm{Re}_\delta \int_0^{y_1/\delta} \frac{U}{U_e} d\left(\frac{y}{\delta}\right) \right] - \frac{1}{U_e} \frac{d}{dx} \left[U_e \mathrm{Re}_\delta \int_0^{y_1/\delta} \left(\frac{U}{U_e}\right)^2 d\left(\frac{y}{\delta}\right) \right]$$

$$= \frac{\tau_w - \tau(y_1/\delta)}{\mu U_e} - \frac{(y_1/\delta)\mathrm{Re}_\delta}{U_e} \frac{dU_e}{dx} \qquad (7\text{-}45)$$

where y_1/δ can have any value from 0 to 1.0. For $y_1/\delta = 1.0$, Eq. (7-45) becomes the usual *momentum integral* equation. For other values of y_1/δ, separate, additional equations are produced. Moses used $y_1/\delta = 1.0$ and 0.3. The latter value corresponds to roughly dividing the momentum loss across the layer in half.

For the required assumption of a velocity profile shape, Moses used an approximate form of the *law of the wake*. He rewrote Eq. (7-10) using the definition of $W(y/\delta)$, Eq. (7-8), and $u^*/U_e = \sqrt{C_f/2}$ and then approximated the *wake function* as

$$W\left(\frac{y}{\delta}\right) = 2\left[3\left(\frac{y}{\delta}\right)^2 - 2\left(\frac{y}{\delta}\right)^3 \right] \qquad (7\text{-}46)$$

The final result is

$$\frac{U}{U_e} = 1 + \frac{\sqrt{C_f/2}}{\kappa} \ln\left(\frac{y}{\delta}\right) - \frac{2\Pi}{\kappa} \sqrt{\frac{C_f}{2}} \left[1 - \tfrac{1}{2} W\left(\frac{y}{\delta}\right) \right] \qquad (7\text{-}47)$$

For the laminar sublayer, Moses used another polynomial, and he notes that the choice of profile for the thin sublayer is not important for integral methods for high-Reynolds-number flows.

A model for the turbulent shear is introduced with an eddy viscosity model [see Eq. (7-36)]. For this integral method, it is only necessary to model the turbulent shear at a single point $y/\delta = y_1/\delta$, since it is only necessary to represent $\tau(y_1/\delta)$ in Eq. (7-45). Moses used

$$\left.\frac{\mu_T + \mu}{\rho U_e \theta}\right|_{y_1/\delta} = 0.0225 + \frac{125}{\mathrm{Re}_\delta} \tag{7-48}$$

based on a direct curve fit of some data by Coles (1962). With all of this, the system consists of two, ordinary differential equations, Eq. (7-45) evaluated at $y_1/\delta = 1.0$ and 0.3, with the velocity profile, Eq. (7-47), [and Eq. (7-46) and the sublayer profile], for two unknowns, $\delta(x)$ (written as Re_δ) and C_f. The equations can be solved numerically using one of the preprogrammed routines for ordinary equations available in any computer center.

Moses (1969) has applied his method to a number of experimental cases, and he has achieved as good agreement as any of the current integral methods.

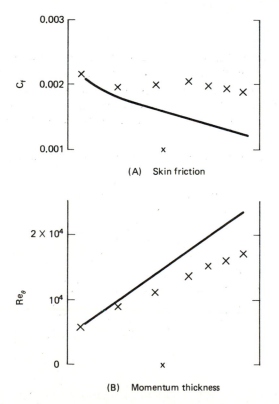

(A)   Skin friction

(B)   Momentum thickness

Figure 7-30 Comparison of the prediction of the integral method of Moses (1969) with the data of Clauser (1956). (From Coles and Hirst, 1969.)

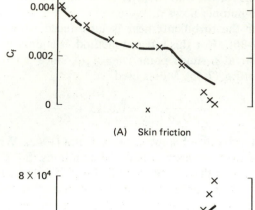

(A)   Skin friction

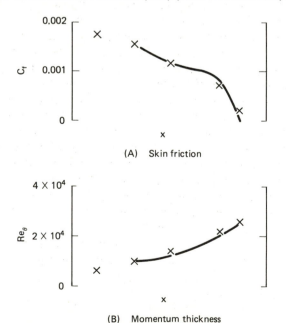

(B)   Momentum thickness

Figure 7-31 Comparison of the prediction of the integral method of Moses (1969) with the data of Schubauer and Klebanoff (1950). (From Coles and Hirst, 1969.)

(A)   Skin friction

(B)   Momentum thickness

Figure 7-32 Comparison of the prediction of the integral method of Moses (1969) with the data of Newman (1951). (From Coles and Hirst, 1969.)

Generally, the agreement is almost as good [for gross quantities such as C_f and $\delta(x)$] as that achieved by the best differential methods to be presented later. In order to display the performance of the various calculation methods, three adverse pressure gradient flows of increasing complexity have been selected for discussion here. All of the methods perform well for flat-plate (zero-pressure-gradient) cases, and the favorable pressure gradient cases are much less challenging. The three cases to be considered are: (1) the equilibrium case 1 $[(\delta^*/\tau_w)(dP/dx) = 2]$ of Clauser (1956) discussed earlier, (2) the classic experiment of Schubauer and Klebanoff (1950) and the airfoil flow of Newman (1951). All three flows are sufficiently well documented to permit unambiguous calculations and comparisons with the data to be made. These flows are three of the more complex that were chosen as test cases for calculation methods at the 1968 AFOSR–IFP–Stanford Conference (see Coles and Hirst, 1969). The predictions of Moses (1969) are compared with the data for these three cases in terms of $\theta(x)$ and $C_f(x)$ in Figs. 7-30 to 7-32. It can be seen that the agreement for the Clauser equilibrium case is poor; it is good for the Schubauer and Klebanoff and Newman cases until separation is approached.

7-8 MEAN FLOW MODELS FOR THE EDDY VISCOSITY AND THE MIXING LENGTH

In order to exploit the availability of computerized methods for solving the differential equations of motion for the mean flow in the boundary layer, it is necessary to have a detailed representation for either the eddy viscosity or the mixing length across the entire layer. Using either Eq. (7-36) or (7-37) (with the required detailed auxiliary representations mentioned above) in Eq. (7-34a), the system of Eqs. (7-30) and (7-34a) can be solved by rather direct extensions of the methods developed in Sec. 4-6. The aims of employing a differential formulation rather than an integral method are to provide better accuracy for C_f, more detailed information [e.g., $U(x, y)$ in the layer] and, hopefully, to work with a more fundamental formulation that requires less of the prior knowledge that is important to the success of integral methods.

It can be expected that the modeling effort will take place in two stages: one for the inner, wall-dominated part of the layer and one for the outer part of the layer. It is easiest to begin with the inner region, since we have, in the *wall law plot* of Fig. 7-4, a compact presentation of empirical information that holds for both zero and nonzero pressure gradients.

7-8-1 Models for the Inner Region

For each region, we begin the modeling effort with the best available experimental information. For the inner region, the *law of the wall* states that

$$u^+ = g(y^+) \qquad (7-49)$$

where the form of $g(y^+)$ is described by the data in Fig. 7-4. Analytical expressions have been derived for two parts of this region. For the sublayer $(0 \le y^+ \le 7)$, there is

$$u^+ = y^+ \tag{7-50}$$

and for the overlap region (roughly, $50 \le y^+ \le 500$) there is

$$u^+ = A \log (y^+) + C = \frac{1}{\kappa} \ln (y^+) + C \tag{7-51}$$

Write the mean flow momentum equation as in Eq. (7-34a) and use $\tau_T \equiv -\rho \overline{u'v'}$ as a more compact notation. Then this equation can be integrated with respect to y using Eq. (7-49) following Coles (1955) to give

$$\frac{\tau + \tau_T}{\tau_w} = 1 + \frac{y}{\tau_w} \frac{dP}{dx} + \frac{v}{u_*^2} \frac{du_*}{dx} \int_0^{y^+} g^2(y^+) \, dy^+ \tag{7-52}$$

Careful study shows that the last two terms can safely be neglected compared to unity up through the entire inner region in virtually all flows. Thus, the equations of motion for the inner region become simply

$$\tau + \tau_T = \mu \frac{\partial U}{\partial y} - \rho \overline{u'v'} = \tau_w \tag{7-53}$$

The simplest part of the inner region to treat is the logarithmic region where $\tau_T \gg \tau$, so Eq. (7-53) becomes

$$\tau_T = \tau_w \tag{7-53a}$$

Using first the *eddy viscosity formulation*, Eq. (7-36), in Eq. (7-53a) results in

$$\mu_T \frac{\partial U}{\partial y} = \tau_w \tag{7-54}$$

The velocity gradient $\partial U/\partial y$ can be evaluated in this region from Eq. (7-51) as

$$\frac{\partial U}{\partial y} = u_* \frac{1}{\kappa y} \tag{7-55}$$

Noting that $u_* \equiv \sqrt{\tau_w/\rho}$ and substituting Eq. (7-55) into Eq. (7-54) produces

$$\mu_T = \kappa \rho u_* y \tag{7-56}$$

which is the desired result—an *eddy viscosity model* for the logarithmic part of the inner region. On reflection, it is clear that one could have probably guessed this result by noting that viscosity = density × velocity × length and picking u_* as the appropriate characteristic velocity and y the appropriate characteristic length scale for the inner region.

This development can be repeated with the *mixing-length formulation*.

One uses Eq. (7-37) in Eq. (7-53a) to obtain

$$\rho \ell_m^2 \left| \frac{\partial U}{\partial y} \right| \frac{\partial U}{\partial y} = \tau_w \tag{7-57}$$

We have $\partial U/\partial y$ in Eq. (7-55). Upon substituting into Eq. (7-57), a *mixing-length model* for the logarithmic part of the inner region emerges as

$$\ell_m = \kappa y \tag{7-58}$$

Note that Eq. (7-58) could have been found from Eq. (7-56) using Eq. (7-40), or vice versa.

The modification of these models for the inner part of the inner region ($y^+ < 50$ to 60) has been accomplished by two workers using two philosophically different approaches. Van Driest (1965a) used *deductive* logic, and Reichardt (1951) used *inductive* logic. Van Driest tried to represent the physics of the damping of the turbulent eddies by the rigid wall by likening that flow to the laminar flow near an oscillating wall in a fluid otherwise at rest. That problem had been solved by Stokes (1851), who showed that the motion diminishes with distance from the wall as $\exp(-y/A)$, where A is a factor depending on the frequency of oscillation and the kinematic viscosity of the fluid. If the plate is fixed, and the fluid oscillates (crudely speaking, as in the eddies), Van Driest reasoned that the factor $[1 - \exp(-y/A)]$ should be applied to the fluid oscillation to obtain the damping effect of the wall. He used the *mixing-length formulation* and took ℓ_m as decreased near the wall by the factor above. Thus, using the model for the logarithmic portion of the inner region, Eq. (7-58), a model for the whole region can be written

$$\tau + \tau_T = \tau_w = \mu \frac{\partial U}{\partial y} + \rho \kappa^2 y^2 \left[1 - \exp\left(\frac{-y}{A} \right) \right]^2 \left| \frac{\partial U}{\partial y} \right| \frac{\partial U}{\partial y} \tag{7-59}$$

or in dimensionless terms as

$$\frac{\tau + \tau_T}{\tau_w} = 1 = \frac{du^+}{dy^+} + \kappa^2 (y^+)^2 \left[1 - \exp\left(\frac{-y^+}{A^+} \right) \right]^2 \left(\frac{du^+}{dy^+} \right)^2 \tag{7-60}$$

It remains now to find the factor A^+, which is a dimensionless *effective frequency* of the turbulent fluctuations. Van Driest (1956a) accomplished that by solving (algebraically) Eq. (7-60) to produce an ordinary differential equation for $u^+(y^+)$,

$$\frac{du^+}{dy^+} = \frac{2}{1 + \sqrt{1 + 4\kappa^2(y^+)^2[1 - \exp(-y^+/A^+)]^2}} \tag{7-61}$$

which can be numerically solved with $u^+(0) = 0$ for $u^+(y^+)$. But we know $u^+(y^+)$; it is defined by the data on Fig. 7-4. Van Driest (1956a) simply tried various values and found that $A^+ = 26$ with $\kappa = 0.40$ produces an excellent representation of the data as shown in Fig. 7-33. For any particular set of wall law constants, A (and hence κ) and C, A^+ changes slightly. With all of this, the

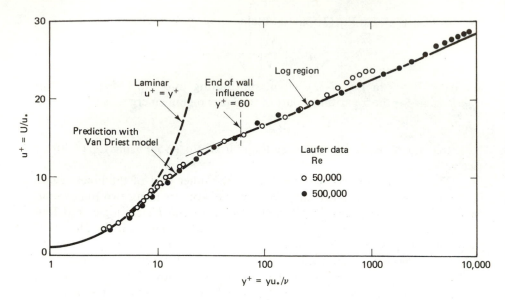

Figure 7-33 Comparison of prediction based on the Van Driest model and experiment for the wall law region. (From Van Driest, 1956a.)

Van Driest *mixing-length model* for the complete inner region is

$$\ell_m = \kappa \left[1 - \exp\left(\frac{-yu_*}{26\nu}\right) \right] y \tag{7-62}$$

An equivalent eddy viscosity model can be found using Eq. (7-40):

$$\mu_T = \rho\kappa^2 \left[1 - \exp\left(\frac{-yu_*}{26\nu}\right) \right]^2 y^2 \left|\frac{\partial U}{\partial y}\right| \tag{7-63}$$

Reichardt (1951) approached the same task quite differently. He simply sought a curve that would mimic the experimentally observed variation of $u^+(y^+)$ between the logarithmic region and the wall, where he reasoned that the eddy viscosity should decay as $\mu_T \sim y^3$ based on the continuity equation. The form selected was

$$\mu_T = \kappa\rho\nu \left[\left(\frac{yu_*}{\nu}\right) - y_a^+ \tanh\left(\frac{yu_*}{\nu y_a^+}\right) \right] \tag{7-64}$$

The constant y_a^+ is chosen to obtain the best fit to the experimental $u^+(y^+)$ in Fig. 7-4. For the Clauser (1956) wall law constants, $y_a^+ = 9.7$ with $\kappa = 0.41$. Again, Eq. (7-40) can be used to yield the corresponding *mixing-length model*.

Despite the fact that the development of these two models was very different, they are to all intents and purposes equal as far as their ability to reproduce the experimentally observed $u^+(y^+)$. That, after all, was the only goal of their development. At the mean flow level of analysis, we regard Fig.

7-4 as representing all we know about turbulent flow. Some workers prefer the use of the Reichardt model in numerical solution procedures because it does not require evaluating $\partial U/\partial y$ as a factor times $\partial U/\partial y$ [compare Eqs. (7-63) and (7-64), noting Eq. (7-36)].

We noted earlier that the major effect of roughness is to shift the logarithmic region of the *law of the wall* down by $\Delta u^+(k^+)$; roughness does not change the slope of $u^+(y^+)$ (see Sec. 7-3-1). One can, therefore, attempt to modify the Van Driest and Reichardt models for roughness. Indeed, Van Driest (1956a) did so by adding an extra term to the wall damping factor in Eq. (7-62):

$$\ell_m = \kappa\left[1 - \exp\left(\frac{-y^+}{26}\right) + \exp\left(\frac{-60y^+}{26k^+}\right)\right]y \qquad (7\text{-}65)$$

He did not, however, check the resulting $u^+(y^+, k^+)$ versus the Hama data collection for $\Delta u^+(k^+)$. Further study has shown that Eq. (7-65) cannot be made to predict that data even if the factor "60" is adjusted (see Schetz and Nerney, 1977). In fact, the Van Driest model has proven very difficult to extend to any more general flows than the smooth, solid wall case Van Driest himself studied.

The Reichardt model was extended to roughness cases by Schetz and Nerney (1977). One can note that the only *free* quantity in the Reichardt model is y_a^+, which is a length scale of the order of the thickness of the laminar sublayer. Now, it is reasonable to expect the thickness of the laminar sublayer, and thus also y_a^+, to decrease with increasing roughness. Schetz and Nerney (1977) have determined the variation $y_a^+(k^+)$ such that the correct shift Δu^+ is produced for the Prandtl–Schlichting uniform sand roughness as an example. The required function y_a^+ is shown in Fig. 7-34. Separate functions, $y_a^+(k^+)$, will have to be developed for other specific types of roughness. That, however, is not difficult as long as $\Delta u^+(k^+)$ is available for the type of roughness of interest.

A number of workers have attempted to extend the Van Driest inner region eddy viscosity model to cases with suction or injection by seeking an appropriate form for $A^+(v_0^+)$. Most have done so *heuristically* without recourse to any law of the wall. It is actually a simple matter to carry out the necessary calculations correctly once a specific form of the *law of the wall* for flows with injection or suction has been chosen. Schetz and Favin (1971) performed this task using the Stevenson law, Eq. (7-18).

Near the wall, the momentum equation can be written

$$\tau + \tau_T = \tau_w + \rho v_w U \qquad (7\text{-}66)$$

or

$$\frac{\tau + \tau_T}{\tau_w} = 1 + v_0^+ u^+ \qquad (7\text{-}66a)$$

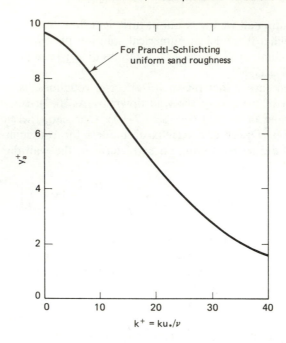

Figure 7-34 Variation of the Reichardt sublayer parameter to account for uniform roughness. (From Schetz and Nerney, 1977.)

This leads to an equation corresponding to Eq. (7-61), the one used by Van Driest to find $A^+(0) = 26$, which is again an ordinary differential equation which can be solved to give $u^+(y^+)$:

$$\frac{du^+}{dy^+} = \frac{2(1 + v_0^+ u^+)}{1 + \sqrt{1 + 4\kappa^2(1 + v_0^+ u^+)(y^+)^2[1 - \exp(-y^+/A^+)]^2}} \qquad (7\text{-}67)$$

with $u^+(0) = 0$. The universal constant, κ, is generally taken as 0.40 to 0.43, and Van Driest used 0.40 to derive the value $A^+(0) = 26$. For a given v_0^+, a trial value of A^+ can be input into Eq. (7-67) and $u^+(y^+)$ can easily be produced numerically. This can be rearranged into appropriate variables and plotted on a Stevenson law plot. A successful value of A^+ for the given v_0^+ is one such that the log portion of the curve falls on that for all other v_0^+'s, including $v_0^+ \equiv 0$.

The results are shown in Fig. 7-35 together with other suggested distributions. A Stevenson law plot of the resulting nondimensional velocity distribution is given in Fig. 7-36, where the excellent correlation achieved may be noted.

It is also not difficult to extend the Reichardt inner region eddy viscosity model to flows with suction or injection as shown by Schetz and Favin (1971). In particular, one needs $(\partial U/\partial y)$ from the wall law in the log region. Fortu-

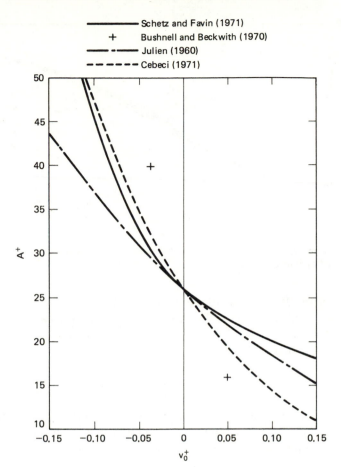

Figure 7-35 Various proposals for the Van Driest damping factor for injection or suction.

nately, both the Stevenson and Simpson laws have the same value for this quantity, so that no choice between these two laws is necessary. Using the momentum equation in the form of Eq. (7-66a) and evaluating $\partial U/\partial y$ in the log region as

$$\frac{\partial U}{\partial y} = \frac{u_*}{y\kappa}(1 + v_0^+ u^+)^{1/2} \tag{7-68}$$

we can write

$$\tau_w(1 + v_0^+ u^+) = \tau_T = \mu_T \frac{\partial U}{\partial y} = \mu_T \frac{u_*}{y\kappa}(1 + v_0^+ u^+)^{1/2} \tag{7-69}$$

This can be rearranged to say that

$$\mu_T = \kappa(\rho u_* y)(1 + v_0^+ u^+)^{1/2} \tag{7-70}$$

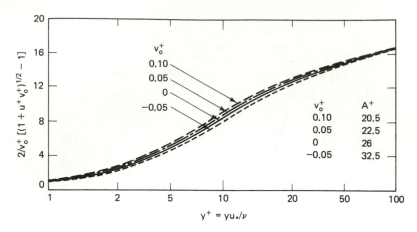

Figure 7-36 Stevenson wall law plot of the predictions of the Van Driest model as extended by Schetz and Favin (1971).

in the log region. But for the additional factor $(1 + v_0^+ u^+)^{1/2}$, this is the same as the earlier result for solid walls. A smooth transition between the required behavior near the wall, $\sim(y^+)^3$, and that in Eq. (7-70) can again be achieved by a hyperbolic tangent. Thus,

$$\mu_T = \kappa \rho v (1 + v_0^+ u^+)^{1/2} \left[y^+ - y_a^+ \tanh\left(\frac{y^+}{y_a^+}\right) \right] \qquad (7\text{-}71)$$

The value of y_a^+ is essentially the value of y^+ at the edge of the laminar sublayer, and this must be known as a function of v_0^+. Stevenson (1963) has suggested a complicated function that can be adequately represented by

$$y_a^+ = \frac{3.65}{v_0^+ + 0.344} \qquad (7\text{-}72)$$

Note, however, that this reduces to $y_a^+(v_0^+ = 0) = 10.7$, which will not match the Clauser wall law constants with $\kappa = 0.41$. This equation will result in a wall law correlation equivalent to that shown in Fig. 7-36 (see Schetz and Favin, 1971).

Neither the Stevenson or Simpson laws nor Eqs. (7-63) (with Fig. 7-35) and (7-71) [with Eq. (7-72)], which are derived from them, attempt to make any distinction between roughness and porosity effects. This was done by Schetz and Nerney (1977) by extending the basic Reichardt model to cases with both roughness and porosity and injection or suction based on their wall law results for a sintered metal surface shown in Fig. 7-9. If one wishes to include blowing as well as roughness, one must look now for $y_a^+(k^+, v_0^+)$ such that the experimental wall law is reproduced. We already have given a complete $y_a^+(k^+, 0)$ for the simple case of uniform sand roughness (see Fig. 7-34).

Since the shift Δu^+ that was observed on the rough, porous sintered

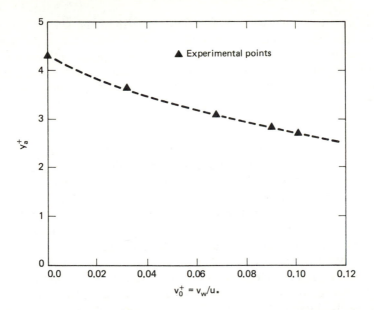

Figure 7-37 Variation of the Reichardt sublayer parameter to reproduce the wall law observed for injection through a sintered metal surface. (From Schetz and Nerney, 1977.)

metal surface was greater than that for uniform sand as reported by Prandtl and Schlichting, the uniform sand curve for $y_a^+(k^+)$ would not reproduce the results. It was found necessary to use $y_a^+ = 4.3$ to produce the observed Δu^+ shift. Thus, a "conjectured" curve through that single point was presented only as a typical example. Separate functions $y_a^+(k^+, 0)$ have to be developed for other specific types of rough, porous surfaces. The experiments were at a single actual roughness size k that produced a nearly constant $k^+ = 7$. Since τ_w, and hence u_*, varies with v_w, the value of k^+ varies with v_0^+ for a constant k. However, the variation of k^+ was only about 20%, which should be negligible. The resulting $y_a^+(7, v_0^+)$ is shown in Fig. 7-37 and good success was achieved in matching the behavior of the experimental wall law for this surface. In order to determine the general function $y_a^+(k^+, v_0^+)$, further injection experiments with other types of porous surfaces will have to be conducted.

7-8-2 Models for the Outer Region

For the outer region, there are again two rather different approaches, but in this case, there is one approach leading to a *mixing-length model* and a different approach leading to an *eddy viscosity model*. A number of workers, including Prandtl himself, have used a simple, physical argument leading to a *mixing-length model*. It is reasoned that the mixing length must be proportional to some length scale on dimensional grounds. Further, in the outer region, the distance from the wall y is less significant than the size of the viscous region δ. Also, the mixing length should be a constant in the outer part of the layer, since the explicit effects of the wall are negligible, and the eddies are insensitive to transverse locations in such a turbulent zone. In this way, one arrives at $\ell_m \sim \delta$. The proportionality constant is order $\frac{1}{10}$, and a value of 0.09 is generally accepted. Thus,

$$\ell_m = 0.09\delta \tag{7-73}$$

This same model, with the same constant, is presumed to apply for all flows either with or without pressure gradients. This model has been used by Spalding and his coworkers at Imperial College in England (see Patankar and Spalding, 1967) with good success reported. Perhaps curiously, this model is not converted into an equivalent *eddy viscosity model* using Eq. (7-40). In fact, most workers, other than the Imperial College group, choose to work with an *eddy viscosity model* developed directly for the outer region.

In light of the discussions of Secs. 7-3-1 and 7-3-3, it should not be a surprise that Clauser (1956) developed an *eddy viscosity model* for the outer region by using his view of the generalized *defect law plot*. It is simplest to discuss the flat-plate case initially. First, Clauser (1956) retained the assumption that the turbulent transport coefficient should be a constant across the outer region, but he used that assumption with an *eddy viscosity* rather than a *mixing length*. Thus, he took

$$\mu_T = \mu_T(x) \neq f(y) \tag{7-74}$$

for the outer region. Second, he used the fact that a turbulent velocity profile *appears* to intersect the wall ($y = 0$) at a nonzero value of the velocity [$U(0) \approx U_e/2$]. Putting these ideas together, he proposed to treat the outer part of the turbulent boundary layer as if it were *pseudolaminar* with some augmented viscosity $\mu_T(x)$ and with a *slip* velocity at $y = 0$. Using

$$\frac{U}{U_e} = g'\left(\frac{y}{\delta}\right) \tag{7-75}$$

in the mean flow momentum equation, there results

$$gg'' + \left(\frac{\mu_T}{\rho U_e \delta(d\delta/dx)}\right)g''' = 0 \tag{7-76}$$

For actual laminar flows, Blasius used $y/\sqrt{x}$, but the results gave $\delta \sim \sqrt{x}$, so y/δ amounts to the same thing. Rescaling both variables to $f(\eta)$, one can obtain

$$f''' + ff'' = 0 \tag{7-76a}$$

which is the same as the equation used by Blasius for the real laminar flat-plate problem (see Sec. 4-3-1). Clauser (1956) solved this equation, but now for various values of the *slip* velocity $U(0)/U_e$ with the results shown in Fig. 7-38.

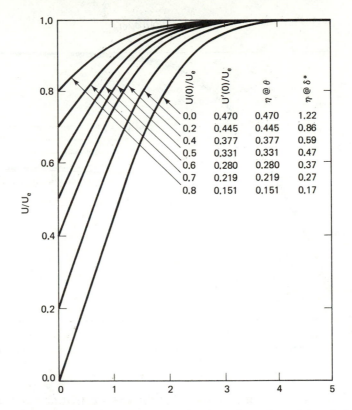

U(0)/U_e	U'(0)/U_e	$\eta @ \theta$	$\eta @ \delta^*$
0.0	0.470	0.470	1.22
0.2	0.445	0.445	0.86
0.4	0.377	0.377	0.59
0.5	0.331	0.331	0.47
0.6	0.280	0.280	0.37
0.7	0.219	0.219	0.27
0.8	0.151	0.151	0.17

Figure 7-38 Solutions of the Blasius equation for various values of *slip* velocity at the wall. (From Clauser, 1956.)

The problem now becomes finding a method to rescale the coordinates so that these profiles will all collapse (or nearly so) into one curve on something resembling a *defect law plot*. In that way, these *pseudolaminar* profiles would mimic the behavior of the real turbulent profiles. Clauser (1956) tried two choices—making the curves coincide at $y/\delta = 0$ and making them have the same slope at $y/\delta = 0$. In both cases, the area above the curve(s) is kept equal. For the second choice, the coordinates are

$$\frac{U - U_e}{U_e \sqrt{(\eta)_{\delta*} \, U'(0)/U_e}} \quad \text{and} \quad \eta \sqrt{\frac{U'(0)/U_e}{(\eta)_{\delta*}}} \qquad (7\text{-}77)$$

where $(\eta)_{\delta*}$ is (η) at $y = \delta^*$, with the result shown in Fig. 7-39, which indeed resembles a *defect law plot*. The process of achieving complete equivalence of the two plots leads to the desired *eddy viscosity* model. Clauser (1956) reasoned on dimensional grounds that $\mu_T \sim$ density × velocity × length. For the characteristic velocity, he made the obvious choice of u_*. For the character-

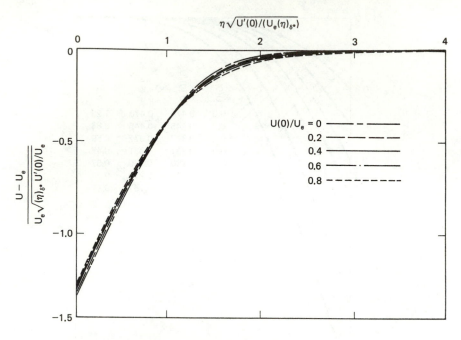

Figure 7-39 Replot of the solutions of the Blasius equation, making the derivatives coincide at $y/\delta = 0$. (From Clauser, 1956.)

istic length, he chose to use a new integral thickness Δ that he developed from the turbulent *defect law plot.*

$$\frac{\Delta}{\delta} \equiv -\int_0^1 \frac{U - U_e}{u_*} \, d\left(\frac{y}{\delta}\right) \tag{7-78}$$

For a flat plate, $\Delta/\delta = 3.6$. Thus,

$$\mu_T = C\rho u_* \Delta \tag{7-79}$$

But, by definition,

$$\Delta = \frac{U_e}{u_*} \delta^* \tag{7-80}$$

[compare Eqs. (7-78) and (2-13)], so the model becomes

$$\mu_T = C\rho U_e \delta^* \tag{7-79a}$$

For the *pseudolaminar* profile in Fig. 7-39,

$$\frac{\tau_w}{\rho} = \frac{\mu_T}{\rho} \frac{\partial U}{\partial y}\bigg|_{y=0} = \frac{\mu_T}{\rho} \frac{U'(0)}{U_e} (\eta)_{\delta*} \frac{U_e}{\delta^*}$$

$$= CU_e^2 \frac{U'(0)}{U_e} (\eta)_{\delta*} \tag{7-81}$$

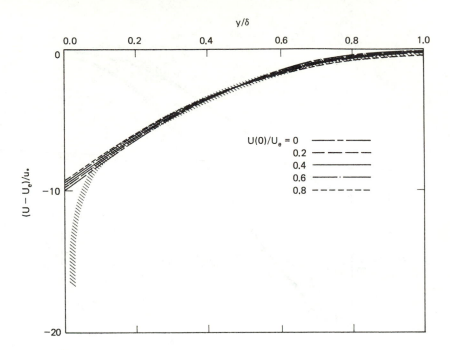

Figure 7-40 Comparison of turbulent data and calculated *pseudolaminar* profiles on a defect law plot. (From Clauser, 1956.)

Thus,

$$u_* \equiv \sqrt{\frac{\tau_w}{\rho}} = U_e \sqrt{C(\eta)_{\delta*} \, U'(0)/U_e} \qquad (7\text{-}82)$$

and

$$\Delta = \frac{\delta^*}{\sqrt{C(\eta)_{\delta*} \, U'(0)/U_e}} \qquad (7\text{-}83)$$

The correct coordinates for a *defect law plot* are $(U - U_e)/u_*$ and y/δ. The *pseudolaminar* coordinates of Eq. (7-77) can be related to these using Eqs. (7-82) and (7-83) as

$$\frac{U - U_e}{U_e \sqrt{(\eta)_{\delta*} \, U'(0)/U_e}} = \sqrt{C} \left(\frac{U - U_e}{u_*} \right)$$

$$\eta \sqrt{\frac{U'(0)/U_e}{(\eta)_{\delta*}}} = \frac{1}{\sqrt{C}} \left(\frac{y}{\Delta} \right) = \frac{1}{3.6\sqrt{C}} \left(\frac{y}{\delta} \right) \qquad (7\text{-}84)$$

Finally then, the constant C can be chosen to make the *pseudolaminar* profiles on Fig. 7-39 *fit* the data on the real turbulent *defect law plot*, Fig. 7-3. Clauser (1956) chose $C = 0.018$ with the results shown in Fig. 7-40. Thus, the well-

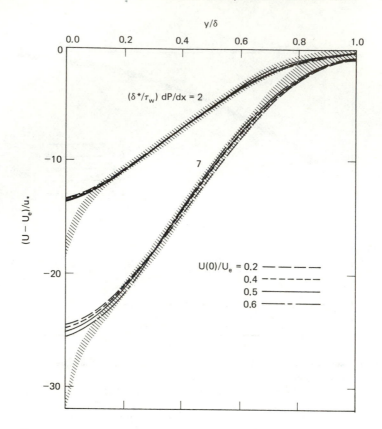

Figure 7-41 Comparison of turbulent data and calculated *pseudolaminar* profiles for equilibrium pressure gradients on a defect law plot. (From Clauser, 1956.)

known *Clauser eddy viscosity model* becomes

$$\mu_T = 0.018\rho U_e \delta^* \qquad (7\text{-}85)$$

Clauser (1956) next went back and established an equivalence between *laminar, wedgetype*, pressure gradient flows and *equilibrium pressure gradient*, turbulent flows. The general procedure is as described above for the flat-plate case. Happily, the same form for the eddy viscosity, Eq. (7-79a) suffices. This may be taken as a demonstration of the soundness of the approach leading to the model. The equivalence achieved is shown in Fig. 7-41. Clauser also found that the constant in Eq. (7-79a) depended weakly, if at all, on $(\delta^*/\tau_w)\,dP/dx$ (see Fig. 7-42). He chose $C = 0.018$ for all cases. Some other workers looking at Fig. 7-42 have since concluded that $C = 0.0168$ is a better choice.

A word of caution is in order here. The Clauser model has been developed only for *equilibrium pressure gradient* flows. If it is applied to flows that

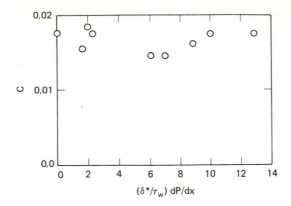

Figure 7-42 Dependence of the proportionality constant in the Clauser eddy viscosity model on the pressure gradient parameter. (From Clauser, 1956.)

are not in equilibrium [i.e., $(\delta^*/\tau_w)\, dP/dx \neq$ constant], one can expect poorer predictions. The model is, however, used very generally without much regard to this basic restriction except to view results for flows that are strongly out of equilibrium with caution. This is mainly because there is no other known eddy viscosity model that is more generally applicable.

It is generally presumed that mean flow models for the outer region are not explicitly influenced by roughness and suction or injection. Those processes directly affect only the inner region.

7-8-3 Composite Model for the Whole Boundary Layer

To make mean flow, turbulent boundary layer calculations, the analyst must choose models for the inner and outer regions—for example, the Reichardt model for a smooth solid surface, Eq. (7-64), for the inner region and the Clauser model, Eq. (7-85), for the outer region. A composite model is easily constructed, since the outer region models are constants across the layer at values much greater than μ and the inner region models all begin with μ_T or ℓ_m at zero (at $y = 0$), which then grow with increasing y [see, for example, Eq. (7-56)]. One simply first calculates the constant μ_T (or ℓ_m) based on the outer region model and then uses the inner model for all values of y (starting up from $y = 0$) where μ_T (or ℓ_m) is less than $(\mu_T)_{outer}$ [or $(\ell_m)_{outer}$]. The outer region model result is used for all other points.

This all may sound complicated, and indeed it is, if done by hand. However, these detailed models are only used within computerized, numerical solution procedures. In such a case, one has a subroutine in the overall code which performs these operations. The computer time, and thus cost, for these calculations is minimal, since only simple arithmetic is involved.

7-9 NUMERICAL SOLUTION METHODS FOR MEAN FLOW FORMULATIONS

At this level of formulation, one is concerned with solving Eqs. (7-30) and (7-34a) with a composite, mean flow turbulent transport model of the type discussed in Sec. 7-8 for essentially arbitrary dP/dx. With one very important exception, this is really no more complicated than the numerical solution of laminar boundary layer problems discussed in Chapters 4 and 5. For the turbulent case, the eddy viscosity, for example, will necessarily vary with y following the models chosen for the inner and outer regions, but in a laminar case the laminar viscosity μ may also vary with y if the temperature varies. In either case, a subroutine is written to calculate the viscosity using the appropriate relations. Indeed, such a subroutine might have a name like VISCOS and be capable of calculating either the laminar viscosity with the Sutherland law if the fluid is air, for example, or calculating an eddy viscosity with one of several choices for models, all depending on some logical switches in the code which respond to the value of parameters in the subroutine calling sequence.

The one special complication in the numerical solution of turbulent boundary layers compared to laminar cases comes directly from the shape of the velocity profile near the wall as has been discussed earlier. Since the velocity changes rapidly over a small distance, it is necessary to have a very fine grid spacing Δy in that region to describe the profile accurately. Consider the profiles in Fig. 7-4 and say that about five grid points are required in the laminar sublayer to describe the profile. The laminar sublayer extends to about $y^+ = 5$, and the boundary layer may extend to $\delta^+ \approx 5000$. In this way, a requirement for roughly 5000 grid points across the boundary layer is apparently derived. This is to be contrasted with laminar cases where about 25 grid points are commonly used. Crudely speaking, computer cost scales with the number of grid points, so a factor of 100 to 200 on cost could be expected.

The resolution of this difficulty is a good example of the value of thinking about the physics of the situation before simply attacking a problem on the computer by brute force. A study of velocity profiles such as those presented earlier in this chapter reveals that while the velocity changes very rapidly near the wall, it changes rather slowly with y over the rest of the profile. Thus, a small Δy (compared to δ) is required near the wall, but a much larger Δy would be adequate in the outer part of the layer. This leads to the idea that a variable, increasing Δy (with increasing y) would be very useful. This results in extra complexity in the numerical analysis of the differential boundary layer equations of motion beyond that discussed in Sec. 4-6, but that is a one-time cost in analysis and programming, and the reduction in computer execution cost per run is so dramatic that the use of a variable grid spacing is compelling on cost grounds. The most common scheme is based on

$$(\Delta y)_{m+1} = k(\Delta y)_m \tag{7-86}$$

with $1.05 \le k \le 1.10$. This is obviously a geometrically increasing sequence.

With this scheme and $k = 1.09$, about 100 grid points across the layer have been found to be satisfactory.

Turbulent boundary layer calculations require the use of a large number of grid points no matter what choice of Δy is used, so efficiency of calculation is always very important. The implicit method of Blottner is strongly recommended here. A very convenient code based on the Blottner method and Eq. (7-86) with many options for the user's easy choice of eddy viscosity models (and even of the constants involved) was developed for the NASA Langley Research Center, by Miner et al. (1975). This code can also treat laminar and/or compressible flows.

To begin to demonstrate the utility of numerical solutions of a mean flow formulation in predicting turbulent boundary layers, some comparisons from the work of Smith et al. (1965) have been chosen. This work employed the Van Driest eddy viscosity model (with $\kappa = 0.40$) for the inner region and the Clauser model for the outer region (with $C = 0.0168$). A so-called *intermittency factor* also multiplied the outer eddy viscosity in an attempt to account for the effects of the intermittency near the outer edge of the layer. The use of such a factor has only a small effect on the results. Figures 7-43 and 7-44 show predictions and experiment for a flat-plate flow on a *wall law plot* and a *defect law plot*. Clearly, excellent agreement was achieved.

For the Stanford competition mentioned earlier, Cebeci and Smith (1969) extended their inner region model in an effort to account for strong pressure

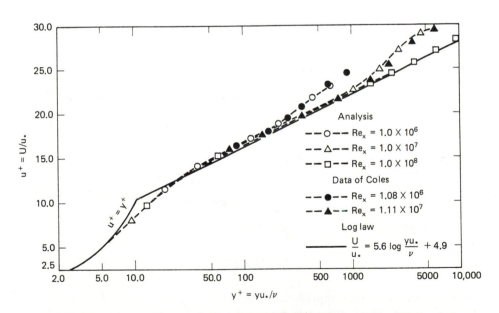

Figure 7-43 Comparison of predictions of Smith et al. (1965) with Van Driest and Clauser eddy viscosity models and data for the wall region.

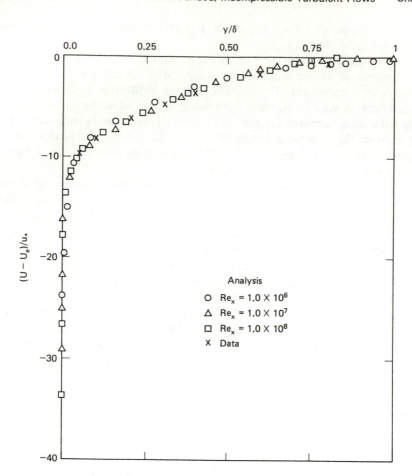

Figure 7-44 Comparison of predictions of Smith et al. (1965) with Van Driest and Clauser eddy viscosity models and data for the outer region of flat-plate boundary layers.

gradients. They proposed modifying the Van Driest damping factor A as

$$A = 26v\left(\frac{\tau_w}{\rho} + \frac{y}{\rho}\frac{dP}{dx}\right)^{-1/2}$$

(7-87)

In later work, they have also suggested that

$$A = 26(v/u_*)\left[1 - 11.8(vU_e/u_*^3)\frac{dU_e}{dx}\right]^{-1/2}$$

(7-88)

One can question this approach, since the *law of the wall* is known to be virtually unaffected by pressure gradients almost to separation, and the Van

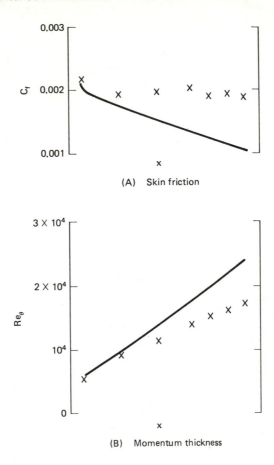

Figure 7-45 Comparison of the prediction of Cebeci and Smith (1969) using a Van Driest–Clauser eddy viscosity model with the data of Clauser (1956). (From Coles and Hirst, 1969.)

Driest model with $A^+ = 26$ faithfully reproduces that law. Cebeci and Smith do not show how their model with Eq. (7-87) reproduces data from flows with strong pressure gradients on a *wall law plot*. Their results for the three test cases chosen for comparisons in this book are given in Fig. 7-45 for the Clauser 1 equilibrium pressure gradient and Figs. 7-46 and 7-47 for the non-equilibrium pressure gradient cases of Schubauer and Klebanoff and Newman. The agreement with the Clauser case data is not as good as one might expect, but we shall see that such is the case for all the analyses. The strong variation of Clauser's equilibrium pressure gradient parameter $(\delta^*/\tau_w)\, dP/dx$ for the Schubauer and Klebanoff case is shown in Fig. 7-48. The agreement between prediction and experiment deteriorates sharply where that parameter begins to

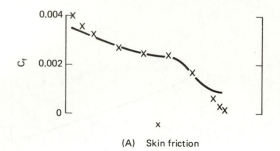

(A) Skin friction

(B) Momentum thickness

Figure 7-46 Comparison of the prediction of Cebeci and Smith (1969) using a Van Driest–Clauser eddy viscosity model with the data of Schubauer and Klebanoff (1950). (From Coles and Hirst, 1969.)

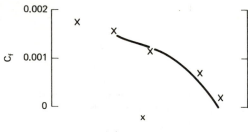

(A) Skin friction

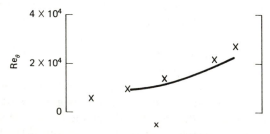

(B) Momentum thickness

Figure 7-47 Comparison of the prediction of Cebeci and Smith (1969) using a Van Driest–Clauser eddy viscosity model with the data of Newman (1951). (From Coles and Hirst, 1969.)

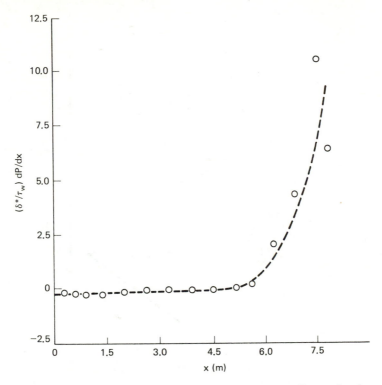

Figure 7-48 Equilibrium pressure gradient parameter versus distance for the experiment of Schubauer and Klebanoff (1950).

vary as could be expected given the restrictions on the Clauser outer region eddy viscosity model used. Looking at all three cases, it might be concluded that the more elaborate, differential method of Cebeci and Smith using a Van Driest–Clauser composite eddy viscosity model is only slightly superior to the Moses integral method for predicting C_f.

At the same level of turbulence modeling, one could use a composite mixing-length model rather than an eddy viscosity approach. The difference would be in the outer region where the statement in Eq. (7-73) is not explicitly restricted to equilibrium pressure gradients. Ng et al. (1969) used such an approach with the Van Driest model (with $\kappa = 0.435$ and $A^+ = 25.3$) for the inner region in the Stanford competition, and their predictions are compared with the three test cases in Figs. 7-49 to 7-51. The performance can be seen to be about equal to that for the eddy viscosity approach.

The last class of flows that will be considered are those with injection. Cebeci has also extended his model to such cases by developing a form for $A^+(v_0^+)$ as described earlier (see Fig. 7-35). There has not been an organized predictor competition for this class of flows, so completely unequivocal comparisons are not possible.

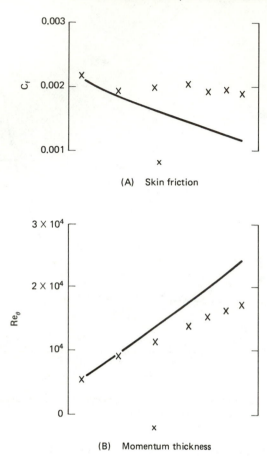

(A)   Skin friction

(B)   Momentum thickness

Figure 7-49   Comparison of the prediction of Ng et al. (1969) using a mixing-length model with the data of Clauser (1956). (From Coles and Hirst, 1969.)

Some comparisons of predictions with data are available from the original works of the individual predictors. A prediction from Cebeci and Mosinskis (1970) is compared with one of the experimental cases of Simpson (1968) in Fig. 7-52. Predictions from Schetz and Favin (1971) using the extended Reichardt model of Eq. (7-71) with (7-72) and two cases from Simpson (1968)—one with injection and one with suction—are shown in Fig. 7-53. The performance of the two formulations can be judged as roughly equal.

Taken as a whole, these comparisons demonstrate a reasonably good capability for predicting turbulent boundary layers with computerized, numerical solutions of mean flow formulations using eddy viscosity or mixing-length models.

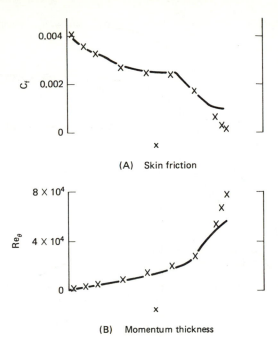

(A)   Skin friction

(B)   Momentum thickness

Figure 7-50   Comparison of the prediction of Ng et al. (1969) using a mixing-length model with the data of Schubauer and Klebanoff (1950). (From Coles and Hirst, 1969.)

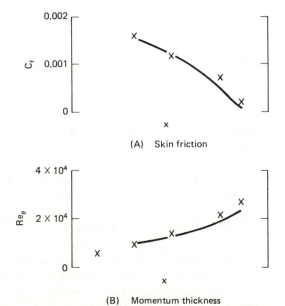

(A)   Skin friction

(B)   Momentum thickness

Figure 7-51   Comparison of the prediction of Ng et al. (1969) using a mixing-length model with the data of Newman (1951). (From Coles and Hirst, 1969.)

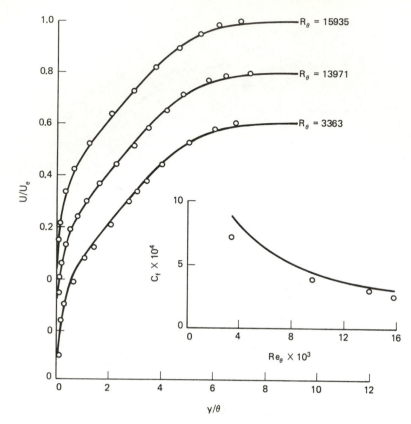

Figure 7-52 Comparison of predictions of Cebeci and Mosinskis (1970) based on an eddy viscosity formulation for flow with injection with the data of Simpson (1968). $v_w/U_e = 0.00784$.

7-10 FORMULATIONS BASED ON TURBULENT KINETIC ENERGY

It is possible to construct turbulent flow formulations at higher levels than those based on the mean properties of the flow alone. If any single quantity is to be selected to represent the fluctuating character of the turbulence, virtually all workers have picked the turbulent kinetic energy (TKE) of the fluctuations defined in Eq. (7-26). There are two general reasons for seeking a formulation that directly involves the fluctuating character of the flow. The first is philosophical. To many people, it just does not seem sensible to analyze a turbulent flow without explicitly treating such a basic feature of turbulence as the fluctuations. Also, one can argue that such a formulation will be at a more *fundamental* level and thus will, presumably, produce better predictions for general flows. The second reason for pursuing this level of formulation is that there are

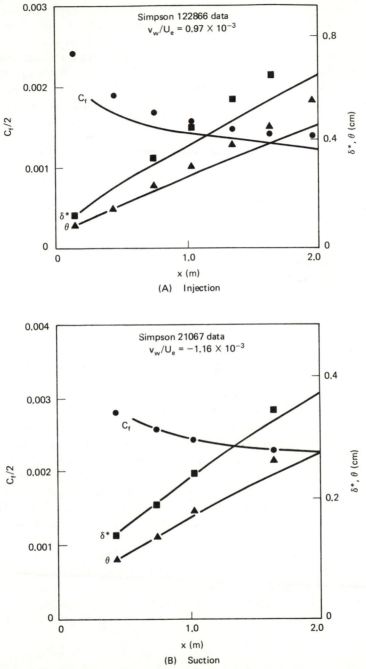

Figure 7-53 Comparison of predictions of Schetz and Favin (1971) using the extended Reichardt eddy viscosity model for cases with injection and suction with the data of Simpson (1968).

some physical situations where a mean flow analysis is clearly inadequate. One easily understood example is a boundary layer flow over a surface with a high level of free-stream turbulence. The boundary layers on the blades of the latter stages of a multistage turbine represent a practical instance where that occurs. Such flows fall outside the data base on which the eddy viscosity or mixing-length models were constructed. There is no way to include the new effects that will result within a mean flow formulation. This reason seems quite compelling, but it is important to note that the occurrence of such flow problems in practice is rather small. For these two general reasons, however, formulations based on the turbulent kinetic energy have been developed to an advanced state.

In order to utilize K, it is necessary to calculate its variation throughout the flow which obviously requires the use of an additional equation, since K is an additional dependent variable. This equation is derived from the Navier–Stokes equations by multiplying each component equation by the corresponding component of the fluctuating velocity, time averaging, summing all three equations, and then making simplifying assumptions as was done with the boundary layer approximation. Under the same restrictions as for the continuity and momentum equations for the turbulent mean flow, the result is

$$\rho\left(U\,\frac{\partial K}{\partial x} + V\,\frac{\partial K}{\partial y} \right) = -\frac{\partial}{\partial y}\,(\overline{\rho v'K'} + \overline{v'p'}) - \rho\overline{u'v'}\,\frac{\partial U}{\partial y} - \mu \sum \overline{\left(\frac{\partial u_i'}{\partial x_j}\right)^2} \qquad (7\text{-}89)$$

    Convection   Diffusion   Production   Dissipation

This type of formulation is usually termed a *one-equation formulation*, since it involves one equation for one *turbulent* quantity K.

In order to implement a turbulent kinetic energy approach, each of the terms on the right-hand side of Eq. (7-89) (or its equivalent in other geometries) must be *modeled* in the same sense that $\overline{u'v'}$ had to be modeled in Eq. (7-34a) via, for example, Eq. (7-36) or (7-37). That is, these terms must be related to the mean flow variables (U, V), the turbulent shear $-\rho\overline{u'v'}$, and/or the turbulence kinetic energy K. This is so, because there are only three equations to determine the three unknowns (U, V, K). Indeed, we see that $\overline{u'v'}$ must still be related to some combination of U, V, and K.

The modeling of the last term on the right-hand side of Eq. (7-89) is generally viewed as noncontroversial. Since viscous dissipation of turbulent energy takes place predominantly at the smaller eddy sizes and these scales have been found to be nearly locally isotropic, the exact result for dissipation under those conditions is generally simply carried over directly, that is,

$$\mu \sum \overline{\left(\frac{\partial u_i'}{\partial x_j}\right)^2} \rightarrow C_D\,\frac{\rho K^{3/2}}{\ell} \qquad (7\text{-}90)$$

where ℓ is a length scale similar to, but not equal to, the mixing length ℓ_m and C_D is an empirical constant of order $\frac{1}{10}$.

Two distinctly different approaches have been proposed for the modeling of the second term on the right-hand side of Eq. (7-89) and the relation of $\overline{u'v'}$ to K. Prandtl (1945) reinvoked the eddy viscosity concept but introduced K through

$$\mu_T = \rho \sqrt{K}\, \ell \qquad (7\text{-}91)$$

This follows directly on observing that viscosity is proportional to density $\times$ velocity $\times$ length. Note that the gradient transport formulation with its limitations remains. Also, one must still model $\ell(y/\delta)$. The second general type of model was introduced by Bradshaw et al. (1967) as

$$\tau_T = -\rho \overline{u'v'} = a_1 \rho K \qquad (7\text{-}92)$$

where $a_1 \approx 0.30$. This is known from experiment (see Fig. 7-20), especially in the inner region. Note that this model does avoid the gradient transport formulation.

The first term on the right-hand side of Eq. (7-89) is generally looked upon as *diffusion*. The modeling of this term is particularly difficult, since good direct data for terms involving the pressure fluctuations out in the flow are not available. By crude analogy with laminar diffusion or, for that matter, with the gradient transport model for the diffusion of momentum, Eq. (7-36), workers who use Eq. (7-91) employ

$$-(\overline{\rho v' K'} + \overline{v'p'}) \rightarrow \frac{\rho \sqrt{K}\ell}{\sigma_K} \frac{\partial K}{\partial y} \qquad (7\text{-}93)$$

where σ_K is a Prandtl number for turbulent kinetic energy, (generally $\sigma_K \approx 1$). This model is called the Prandtl energy method. Bradshaw et al. (1967) proposed

$$(\overline{\rho v' K'} + \overline{v'p'}) = \tau_T \sqrt{\frac{\tau_{\max}}{\rho}}\, G\!\left(\frac{y}{\delta}\right) \qquad (7\text{-}94)$$

where $G(y/\delta)$ is a universal graphed function determined by recourse to flat-plate flow data. This model avoids the gradient transport formulation, but it still requires that $\ell(y/\delta)$ and $G(y/\delta)$ be modeled. A suggested variation for $\ell(y/\delta)$ was also determined from flat-plate flow data, and it is presented as another graph. Last, their shear stress model appears to be valid mainly for external, wall-bounded flows.

For the Prandtl energy method, the modeled form of the TKE equation becomes

$$\rho\!\left(U\frac{\partial K}{\partial x} + V\frac{\partial K}{\partial y}\right) = \frac{\partial}{\partial y}\!\left(\frac{\rho \sqrt{K}\ell}{\sigma_K}\frac{\partial K}{\partial y}\right) + \rho \sqrt{K}\ell\!\left(\frac{\partial U}{\partial y}\right)^2 - C_D\frac{\rho K^{3/2}}{\ell} \qquad (7\text{-}95)$$

With a model for $\ell(y/\delta)$, it is clearly possible to solve Eq. (7-95) using the numerical methods developed for the momentum equation, since the forms are so similar. However, with Bradshaw's modeling, the TKE equation takes on a *hyperpolic*, not *parabolic*, form, and the usual methods for the boundary layer equations do not apply. Bradshaw developed suitable methods, and the interested reader should refer to the original paper for full details. Bradshaw has a convenient computer code which can easily and cheaply be obtained from him.

Just as for the eddy viscosity or mixing length, the modeling at this level is completed by looking at limiting cases of the equations of motion. Near a rigid wall, *convection* and *diffusion* of K are negligible compared to *production* and *dissipation* (see Fig. 7-24), so Eq. (7-95) becomes

$$\rho\sqrt{K}\ell\left(\frac{\partial U}{\partial y}\right)^2 = C_D\frac{\rho K^{3/2}}{\ell} \tag{7-95a}$$

Noting Eqs. (7-36) and (7-91), Eq. (7-95a) can be rewritten as

$$\tau_T^2 = \left(\mu_T\frac{\partial U}{\partial y}\right)^2 = C_D\rho^2 K^2 \tag{7-95b}$$

or

$$\tau_T = C_D^{1/2}\rho K \tag{7-96}$$

Comparing with Eq. (7-92), this gives $C_D = 0.08$ for $a_1 = 0.30$. A little more algebra gives

$$\tau_T = C_D^{-1/2}\rho\ell^2\left(\frac{\partial U}{\partial y}\right)^2 \tag{7-97}$$

This relation has good news and bad news. The good news is that comparing with Eq. (7-37), there results

$$C_D^{-1/4}\ell = \ell_m \tag{7-98}$$

so that any good model for $\ell_m(y/\delta)$ can be used to give an equivalent model for $\ell(y/\delta)$ which was needed. The bad news is that this shows that the Prandtl energy method and the mixing-length formulation collapse to the same thing in the all-important wall region. Note also that the analysis of the wall region above, especially that leading to Eq. (7-96), shows that the Bradshaw TKE formulation also is closely related to the Prandtl energy method in the inner region. Thus, the mixing-length model, the eddy viscosity models (Van Driest or Reichardt), the Prandtl energy method, and the Bradshaw TKE method are all essentially equivalent in the inner region for flows over smooth, solid surfaces with moderate pressure gradients. One can, therefore, expect that any differences in performance in comparisons with data will be due to greater generality (no restriction to equilibrium pressure gradients) in the treatment of

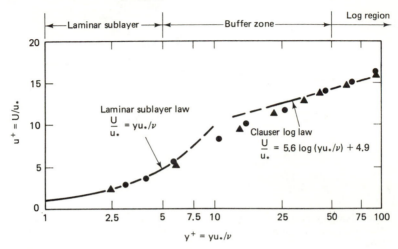

Laminar sublayer ⟶ ⟵ Buffer zone ⟶ Log region

Laminar sublayer law
$$\frac{U}{u_*} = yu_*/\nu$$

Clauser log law
$$\frac{U}{u_*} = 5.6 \log{(yu_*/\nu)} + 4.9$$

$u^+ = U/u_*$

$y^+ = yu_*/\nu$

Figure 7-54 Comparison of predictions with an eddy viscosity formulation and the Prandtl energy method with data correlations for the inner region.

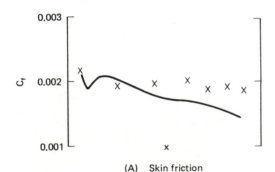

(A) Skin friction

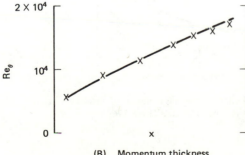

(B) Momentum thickness

Figure 7-55 Comparison of the prediction of the TKE model of Bradshaw et al. (1967) with the data of Clauser (1956). (From Coles and Hirst, 1969.)

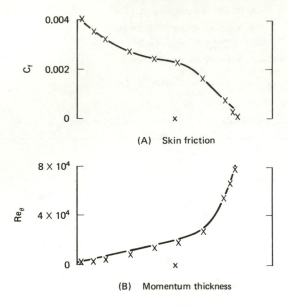

Figure 7-56 Comparison of the prediction of the TKE model of Bradshaw et al. (1967) with the data of Schubauer and Klebanoff (1950). (From Coles and Hirst, 1969.)

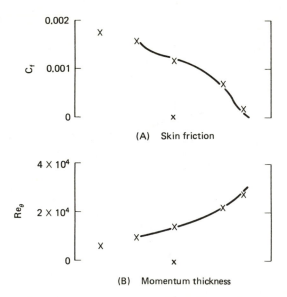

Figure 7-57 Comparison of the prediction of the TKE model of Bradshaw et al. (1967) with the data of Newman (1951). (From Coles and Hirst, 1969.)

the outer region and/or better modeling of the flow as separation is approached.

For flat-plate flow, predictions for the inner region based on the Prandtl energy method are indistinguishable from those based on a Reichardt–Clauser eddy viscosity model as shown in Fig. 7-54 from the author's work. Comparisons of predictions using the Bradshaw TKE method and the three pressure gradient test cases are shown in Figs. 7-55 to 7-57. Generally better agreement with the data is demonstrated than for any of the methods discussed previously. Calculations by the author with the Prandtl energy method show essentially equivalent performance.

There is also the practical advantage of the TKE methods that they can treat flows where the turbulence field does not have the same relation to the mean flow field as for the usual cases used in the data base for the mean flow formulations. Again, we cite the case of flows with high freestream turbulence as, perhaps, the most common example. The author (see Schetz, et al., 1981) has made calculations using the Prandtl energy method for comparison with the results of Huffman et al. (1972) for the experimental situation of a flat plate where the free-stream turbulence level was varied over a rather wide range. A comparison of predictions and experiment is shown in Fig. 7-58. It appears

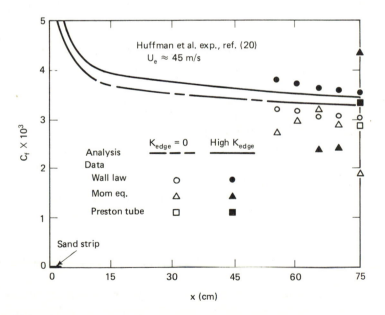

Figure 7-58 Comparison of prediction with the Prandtl energy method and the data of Huffman *et al.* (1972) for low-speed flow with high free-stream turbulence.

that the TKE analysis underpredicts somewhat the effects of increasing K_e, although there is considerable uncertainty in the data. Some of the discrepancy can also be traced to the fact that the experiment involved a leading edge sand strip to trip the flow, for which the analysis cannot account.

There has been much less work aimed at extending TKE models to cases with roughness and suction or injection than for the mean flow models. Bradshaw reports such extensions for his model, but no comparisons with data have appeared in the open literature.

7-11 FORMULATIONS BASED ON TURBULENT KINETIC ENERGY AND A LENGTH SCALE

Many of the deficiencies in the performance of the TKE models have been attributed to the manner in which the length scale is modeled. That quantity generally appears directly in the modeled form of the various terms, for example, Eqs. (7-90), (7-91), and (7-93). At the level of the TKE models, the most that can be accommodated is an algebraic formula, usually taken as $\ell \sim y$ or δ. There are some physical situations where the simple relation of the length scale to some distance or thickness is not clear. Perhaps the easiest case to understand occurs in a free turbulent flow—the merging of the mixing zones from several parallel, coaxial jets, all exhausting in one cross-sectional plane. Near the point of injection, each jet is unaware of the others, and the mixing length can be simply related to the local width of the mixing zone of a single jet. Further downstream, the several mixing zones have grown to the point where they merge, more or less abruptly. The *width* of the mixing zone has now suddenly grown much larger. If there are 50 jets, it becomes 50 times larger. Clearly, the local flow in one of the middle jets cannot instantly respond to the presence of all the other jets in a linear way, as would be implied by taking ℓ as proportional to the new mixing zone width. We must have something better, and one approach has been to seek an independent equation for ℓ or for a new quantity that is a combination of K and ℓ, say, $Z \equiv K^m \ell^n$. Such an equation can also be found by manipulating the Navier–Stokes equations, and the result, under the same restrictions as before and following some modeling, is

$$\rho \left(U \frac{\partial Z}{\partial x} + V \frac{\partial Z}{\partial y} \right) = \frac{\partial}{\partial y} \left(\frac{\mu_T}{\sigma_Z} \frac{\partial Z}{\partial y} \right) + Z \left[C_1 \frac{\mu_T}{K} \left(\frac{\partial U}{\partial y} \right)^2 - C_2 \frac{\rho^2 K}{\mu_T} \right] + S_Z \quad (7\text{-}99)$$

Here σ_Z is a *Prandtl number* for diffusion of Z, S_Z are secondary source terms that appear in some models, and C_1 and C_2 are constants. This equation is almost identical in form to some of the modeled forms of the TKE equation. One still works with the mean flow equations and the turbulent kinetic energy (TKE) equation along with the new Z equation, and this approach has been named a *two-equation model*.

The most active proponents by far of the two-equation models have been Spalding and his coworkers from the Imperial College group, [see the book by Launder and Spalding (1972)]. They have concentrated mainly on the choice $m = \frac{3}{2}$, $n = -1$, giving $Z = K^{3/2}/\ell$, which is proportional to the dissipation rate ε. These models are, therefore, often termed $K\varepsilon$ models. Each term on the right-hand side of the Z (now ε) equation still had to be modeled. This was accomplished largely by analogy with the modeling of Reynolds stress and the TKE equation. For example, diffusion of Z is written as

$$- \overline{\rho v' Z'} \sim \mu_T \frac{\partial Z}{\partial y} \tag{7-100}$$

There are genuine conceptual problems here. If one looks closely at the various terms of the Z equation, one is really modeling terms that can be described as *dissipation of dissipation*, for example. In spite of such philosophical problems, these approaches are aimed at solving a real problem—better specification of the length scale, and they do offer some practical advantages.

The value of C_2 is found by considering the simple case of the decay of turbulence in uniform (in the transverse direction) flow behind a screen. In that situation, one can neglect *production* and *diffusion* and Eq. (7-95) reduces to

$$\rho U \frac{dK}{dx} = -C_D \frac{\rho K^{3/2}}{\ell} = -C_D \frac{\rho^2 K^2}{\mu_T} \sim x^{-2} \tag{7-101}$$

where the x^{-2} decay is known from experiment. Equation (7-99) reduces to

$$\rho U \frac{dZ}{dx} = -C_2 \frac{\rho^2 K Z}{\mu_T} \tag{7-102}$$

Thus

$$C_2 = C_D\left(m - \frac{n}{2}\right) \tag{7-103}$$

To set the value of C_1, it is again helpful to consider the logarithmic portion of the wall region where we have Eq. (7-96) and

$$\frac{\partial U}{\partial y} = \frac{u_*}{\kappa y} = \frac{\sqrt{\tau/\rho}}{\kappa y} \tag{7-104}$$

which comes from Eq. (7-55) and the assumption $\tau = \tau_w$. In this region convection of Z can be neglected, and Eq. (7-99) reduces to give

$$C_1 = \frac{C_2}{C_D} - \frac{\kappa^2 n^2}{\sigma_Z C_D^{1/2}} \tag{7-105}$$

Thus, for $Z = (K^{3/2}/\ell) \sim \varepsilon$, one has $C_1 \approx 1.5$, $C_2 \approx 0.18$, $\sigma_Z \approx 1.0$, and $\kappa \approx 0.41$.

Ng and Spalding (1970) used a two-equation model with K and $K\ell$

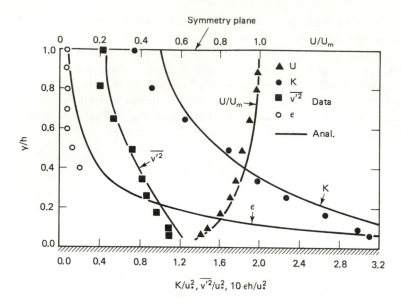

Figure 7-59 Comparison of predictions with a $K\varepsilon$ model by Hossain (1979) and the data of Laufer (1951) for fully developed flow in a channel.

($m = n = 1$) as the two variables. They presented calculations for many of the cases from the Stanford competition and found that the results were almost identical to their earlier results with the mixing-length model (Ng et al., 1969) except for strongly nonequilibrium pressure gradients. In those cases, the two-equation model performed better when compared to experiment.

The $K\varepsilon$ formulation has been widely used, especially for free turbulent flows. A comparison of a prediction using the $K\varepsilon$ model from Hossain (1979) and experiment from Laufer (1951) for fully developed flow in a channel of half-height h is shown in Fig. 7-59. See the book by Launder and Spalding (1972) and the clear exposition by Rodi (1975) for further comparisons with data and additional details. There have been no sustained attempts to use $K\varepsilon$ models for flows with roughness and suction or injection.

7-12 FORMULATIONS BASED DIRECTLY ON THE REYNOLDS STRESS

The TKE and $K\ell$ models described above both suffer from the gradient transport concept restriction. The eddy viscosity (or mixing-length) notions combined with the gradient transport relationship, in general, imply too direct a connection between the mean flow field and the turbulent stress. This remains a limitation on the models that are more advanced than the mean flow models, even though other facets of the turbulence have been included. A way around this problem is to formulate a transport equation for the Reynolds stress, $-\rho \overline{u'v'}$, itself. One is then attacking directly the term that caused the central problem at the lowest level of analysis in Eq. (7-34a). With our usual restrictions the equation for the Reynolds stress is

$$U \frac{\partial \overline{u'v'}}{\partial x} + V \frac{\partial \overline{u'v'}}{\partial y} = -\overline{v'^2} \frac{\partial U}{\partial y} - \frac{\partial}{\partial y}\left(\overline{v'u'v'} + \frac{\overline{p'u'}}{\rho}\right)$$

$$\qquad\text{Convection}\qquad\qquad\text{Production}\qquad\qquad\text{Diffusion}$$

$$+ \frac{\overline{p'}}{\rho}\left(\frac{\partial u'}{\partial y} + \frac{\partial v'}{\partial x}\right) - 2v\sum\overline{\left(\frac{\partial u'}{\partial x_j}\frac{\partial v'}{\partial x_j}\right)} \qquad (7\text{-}106)$$

$$\qquad\qquad\text{Redistribution}\qquad\qquad\text{Viscous Dissipation}$$

[See the book by Launder and Spalding (1972) for details.]

Of course, Eq. (7-106) contains higher-order turbulence terms [e.g., $\overline{v'(u'v')}$], which must now be treated. It is possible to envision writing a new, separate equation for each term of that type. Such equations will, however, necessarily involve terms of the next higher order. The process could be continued, but the whole scheme is unbounded, since the last equation will always contain terms of a higher order. It is necessary, therefore, to *close* the problem by terminating the sequence at some level. At the mean flow model level, the problem is closed by an eddy viscosity or mixing-length model. At the TKE and $K\ell$ model level, closure is achieved by modeling the higher-order terms in the TKE and Z equations. In like manner, it will now be necessary to model the unknown terms in Eq. (7-106) for the Reynolds stress. Most workers have also found it desirable to add other equations to the system at this level. Equations for ℓ and/or the normal stresses $\overline{u'^2}$, and so on, are common choices.

Although it was not the first model from a chronological point of view, the model of Hanjalic and Launder (1972) follows most simply in logical order from the previous discussion. It is, therefore, presented here first. The *production* term in Eq. (7-106) is modeled by asserting that $\overline{v'^2} \propto K$. The *diffusion* term is modeled through the generalized gradient transport framework, that is,

$$-\frac{\partial}{\partial y}\left(\overline{v'u'v'} + \frac{\overline{p'u'}}{\rho}\right) \sim \frac{\partial}{\partial y}\left(\sqrt{K}\ell\, \frac{\partial \overline{u'v'}}{\partial y}\right) \qquad (7\text{-}107)$$

Note that here a turbulent viscosity ($\mu_T \sim K^{1/2}\ell$) has crept back into the formulation. The *dissipation* term was neglected with respect to the other terms. Finally, the *redistribution* term is modeled via two processes: one with turbulent interactions alone, $\sim(K^{1/2}/\ell)\overline{u'v'}$, and one that arises via the mean flow gradient, $\sim K\,\partial U/\partial y$. The final form of the equation for the Reynolds stress becomes

$$U \frac{\partial \overline{u'v'}}{\partial x} + V \frac{\partial \overline{u'v'}}{\partial y} = \frac{1}{\rho}\frac{\partial}{\partial y}\left(\frac{\mu_T}{\sigma_\tau}\frac{\partial \overline{u'v'}}{\partial y}\right) - C_\tau\left(K\frac{\partial U}{\partial y} + \frac{K^{1/2}}{\ell}(\overline{u'v'})\right) \qquad (7\text{-}108)$$

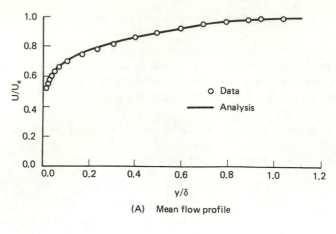

(A)   Mean flow profile

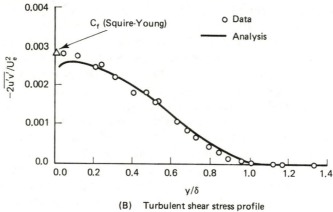

(B)   Turbulent shear stress profile

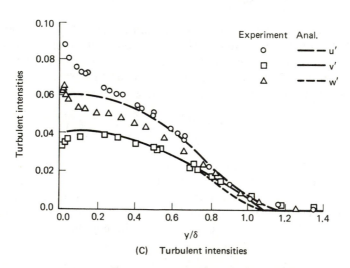

(C)   Turbulent intensities

Figure 7-60   Comparison of predictions with the Donaldson Reynolds stress model by Rubesin et al. (1977) and the flat-plate data of Klebanoff (1954).

with $\sigma_\tau \approx 0.9$ and $C_\tau \approx 2.8$. Since this equation contains K and ℓ, the system is completed with equations for K and ℓ as modeled by the Imperial College group except that τ_T is left as $-\rho \overline{u'v'}$ as the various terms in Eqs. (7-89) and (7-99) are approximated. Note that in the wall region where *convection* and *diffusion* may be neglected compared to the other terms Equation (7-108) collapses to

$$0 = -C_\tau \left(K \frac{\partial U}{\partial y} + \frac{K^{1/2}}{\ell} \overline{u'v'} \right) \tag{7-108a}$$

or

$$\rho \sqrt{K\ell} \, \frac{\partial U}{\partial y} = -\rho \overline{u'v'} \tag{7-108b}$$

which is the *Prandtl energy method* and which itself was shown (see Sec. 7-10) to be equivalent to the mixing-length model! Thus, we should not expect any large improvement in the predictions from this model for wall-bounded flows of the ordinary type.

At a slightly higher level of mathematical complexity, Donaldson (1971) chose to employ an equation for the Reynolds stress, algebraic relations for the length scale ℓ, and separate equations for the normal stresses $\overline{u'^2}, \overline{v'^2}, \overline{w'^2}$. Since each normal stress is found separately, an equation for one-half their sum (i.e., K) is not needed. The details of the modeling of all the terms can be found in the paper. It is worth noting that the turbulent viscosity concept is not invoked in any form. The length scale is, however, not determined from a differential equation. Some comparisons of predictions by Rubesin et al. (1977) based on the Donaldson model and experimental results for a flat boundary layer are shown in Fig. 7-60. The mean flow, C_f, $\overline{u'v'}$, and $\overline{v'^2}$ are well predicted, but $\overline{u'^2}$ and $\overline{w'^2}$ are poorly predicted. The fact that the mean flow and the Reynolds stress can be accurately predicted with a formulation that gives poor predictions for $\overline{u'^2}$ and $\overline{w'^2}$ can be interpreted to indicate that it is perhaps not worthwhile to solve separate equations for $\overline{u'^2}, \overline{v'^2},$ and $\overline{w'^2}$.

The model of Daly and Harlow (1970) employs differential equations for the Reynolds stress, each of the normal stresses and ℓ (via ε), so this model entails the solution of five equations for turbulent quantities in addition to two (or more) mean flow equations. The modeling of the terms in the equations for the Reynolds and normal stresses is similar to that used by Hanjalic and Launder.

7-13 DIRECT TURBULENCE FORMULATIONS

To this point, we have not made any connection with the large body knowledge that has grown in what might be called *statistical turbule* search. Only recently, however, have tentative connections between th

tical theories and the needs of the engineer been made. Here, we shall call such models *direct turbulence formulations*. The basic problem with attempting to treat the fluctuating turbulent flow directly is that such flows are characterized by a range of the excited scales of motion over several orders of magnitude (see Fig. 7-22). Even the most optimistic projections for the capacity of future computing machines fall far short of that estimated to meet this need. Only at low-Reynolds-number conditions does *all-scale* resolution of the turbulence seem at all likely in the foreseeable future. One promising method that has been proposed for alleviating this problem is to limit the attempt at direct, three-dimensional, unsteady treatment of the turbulence to only those scales above a certain size. All the scales that cannot be resolved are then modeled as a *subgrid turbulence* using an eddy viscosity or other transport approximation. The hope underlying this approach is that the smaller-scale structure of the turbulence is nearly universal, so that accurate resolution is not required. The large eddy structure that is presumed to contain the part of the turbulence that changes markedly from flow to flow or condition to condition is treated directly.

7-14 BOUNDARY AND INITIAL CONDITIONS FOR HIGHER-ORDER MODELS

An important point concerning the use of TKE, $K\ell$ or Reynolds stress models emerges. With the additional flow variable (e.g., K) and an accompanying differential equation(s) one needs boundary conditions on that variable(s) as well as the usual requirements for (U, V). This is often not a trivial matter. In some experimental studies, no data for K, or any other turbulence data, are reported. It therefore becomes necessary to *estimate* the values of the initial and boundary conditions. Obviously, this is a perilous undertaking, and the resulting predictions can be expected to be sensitive to the specific choices made. The situation is most difficult with regard to the initial conditions. One must interpret the initial profiles for U (i.e., mean flow variables) to generate an initial profile for K and/or ε and/or $\overline{u'v'}$. Generally, the boundary condition(s) at the outer edge of the mixing region is taken as zero or the free-stream turbulence level, if that is provided. This difficulty is an important limitation on the routine use of these models for preliminary design estimates. In that situation, the actual initial and boundary conditions for the mean flow variables are generally not precisely known and must be estimated themselves. Carrying that process further to produce an initial profile for K, for example, is clearly fraught with difficulties.

PROBLEMS

7.1. Air at standard temperature and pressure (STP) is flowing over a flat plate at 30 m/s. What is the value of C_f at $x = 1$ m? Calculate u_*.

7.2. For the flow in Prob. 7.1, what is the thickness of the laminar sublayer? Estimate the distance to the inner and outer edges of the logarithmic region.

7.3. CO_2 at STP flows over a flat plate at 10 m/s. What is the value of C_f for a smooth plate at $x = 1.0$ m? What is the *allowable* roughness size for *smooth*? What roughness size corresponds to *fully* rough?

7.4. For water at 25°C flowing in a pipe 0.5 m in diameter at 2 m³/s, what is the value of λ for a smooth wall? What is the corresponding value for a concrete pipe? ($k \approx 0.002$ m.)

7.5. A flat plate is immersed in a 100-m/s air stream (at STP). Plot the Reynolds stress versus y at a distance 1 m from the leading edge.

7.6. To test your knowledge of the first part of Chapter 7, try *without looking back at the text* to state the restrictions and limits on region(s) of applicability, if any, on the following items.
(a) $u^+ = g(y^+)$
(b) $u(x, y, z, t) = U(x, y) + u'(x, y, z, t)$
(c) $\ell_m = \kappa y$
(d) $(U - U_e)/u_* = f(y/\delta)$
(e) $\tau_T = -\rho \overline{u'v'}$
(f) $U/u_* = A \log (yu_*/\nu) + C$
(g) $\Delta U/u_* = F(k^+)$
(h) $\mu_T = 0.018\rho U_e \delta^*$

7.7. We have water flow over a surface at a location where $\delta = 2$ cm and the edge velocity is 10 m/s. The local skin friction coefficient is $C_f = 0.002$. What is the value of the eddy viscosity and the mixing length at $y = \delta/1000$? At $\delta/20$? What is the turbulent shear at the same points? What are the eddy viscosity and the turbulent shear at $y = \delta/2$?

7.8. It is accepted as a useful approximation that

$$\frac{U}{U_e} = \left(\frac{y}{\delta}\right)^{1/n}$$

for flat-plate turbulent boundary layers, but for low Re_x, $n \approx 6$ and for very high Re_x, $n \approx 8$. How would the predicted growth rate of the boundary layer thickness be influenced? How do the shapes of the profile differ?

7.9. Consider two airflows over solid surfaces where the boundary layer thickness, δ, is the same and U_e is the same but one is laminar and one is turbulent. (*Note:* This could happen for a Reynolds number near transition and two different background disturbance levels.) What is the ratio of the shear at $y = \delta/2$ in the two cases? You may neglect pressure gradient effects.

7.10. Show that for the Reichardt model, Eq. (7-64), the eddy viscosity behaves as y^3 near the wall, $y = 0$.

7.11. Consider water flow over a flat plate at 8 m/s. At a location 1 m from the leading edge, what is δ, δ^*, θ, C_f, and u_*? What is the thickness of the laminar sublayer? What is the eddy viscosity at a distance $\delta/2$ above the plate? What is the mixing length at the same point?

7.12. Water at 25°C is flowing over a flat plate at 5 m/s. What is the value of the TKE at $x = 0.5$ m and $y = \delta/2$? What is the value of the length scale ℓ? What is the value of the eddy viscosity?

8

FREE TURBULENT SHEAR FLOWS— JETS AND WAKES

8-1 INTRODUCTION

Here we shall be concerned with those turbulent flows where there is no rigid surface present in the region of interest. The two general representatives of this class of flows are wakes and jets, as illustrated schematically in Fig. 8-1. Attention is focused on the flow downstream of the body producing the wake or the walls that initially bound the jet. The major simplifying feature of such flows is that the treatment of an inner, wall-dominated region is eliminated. Thus, one is only concerned with the equivalent of the outer region of a wall-bounded turbulent layer.

Both wake and jet flows can be viewed as consisting of two or more regions. The *near wake* right behind the body and the *potential core* near the injection station of the jet are relatively complex, and they depend on the details of the body shape or the jet nozzle. In the idealized case of small boundary layers both on the outside and the inside of the jet injector, there is a substantial initial region before the developing shear layers from the edges of the jet merge on the axis. This is called the *potential core*, since the flow of the jet fluid in this region is uniform and inviscid. If the flow coming out of the injector is fully developed, or nearly so, there is little or no potential core.

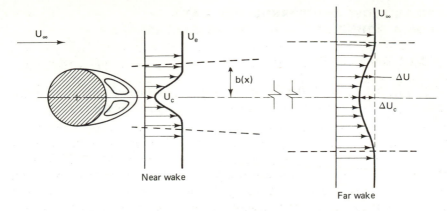

(A)   Typical wake flow

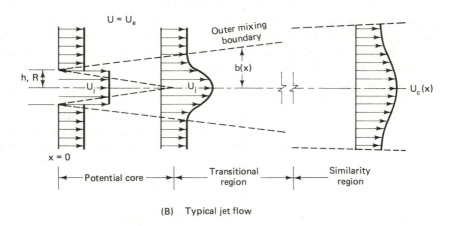

(B)   Typical jet flow

Figure 8-1   Schematic illustrations of free turbulent shear flows.

Harsha (1971) has given the following data correlation for air–air, axisymmetric jets:

$$\frac{x_c}{D} = 2.13(\mathrm{Re}_D)^{0.097} \tag{8-1}$$

which can be used for estimating purposes.

Far behind the body or the jet nozzle, the flows assume a more universal
behavior, and analysis is simpler. A transitional region between the near and
far fields is often also envisioned.

To limit the scope of this chapter to a manageable level, the case of a
round jet in a co-flowing external stream has been chosen as a representative
to illustrate the nature of this class of flows. A more general treatment of the
subject can be found in the monograph by the author (Schetz, 1980).

8-2 MEAN FLOW AND TURBULENCE DATA
FOR A ROUND JET IN A MOVING STREAM

8-2-1 Constant-Density Flows

In Fig. 8-2 we show the variation of the centerline velocity in terms of $U_e/[U_c(x/D) - U_e]$ and a characteristic width $r_{1/2}$ defined by the relation

$$U(r_{1/2}, x) = \frac{U_c(x) + U_e}{2} \tag{8-2}$$

with axial distance. This latter quantity is frequently called the half-radius, and it can be more easily determined accurately than some ill-defined total width where the mixing zone merges asymptotically into the external stream. First, it can be seen that the influence of the parameter m $(\equiv U_e/U_j)$ on the flow field is quite profound. Second, the total distance to final decay where $U_c \rightarrow U_e$ is very long when measured in terms of jet diameters. Nondimensional radial velocity profiles are given in Fig. 8-3 for various x/D. These data indicate that the profiles are apparently *similar* for $x/D \geq 40$, where the term *similar* means that the profiles, expressed in terms of coordinates such as those on Fig. 8-3, remain unchanged with x/D.

Some data of Antonia and Bilger (1973) for the axial turbulence intensity are given in Figs. 8-4 and 8-5. For free turbulent flows, the turbulence intensity is usually normalized with the mean flow velocity difference across the layer ΔU_c rather than U_e or U. Note that the average fluctuations are a substantial

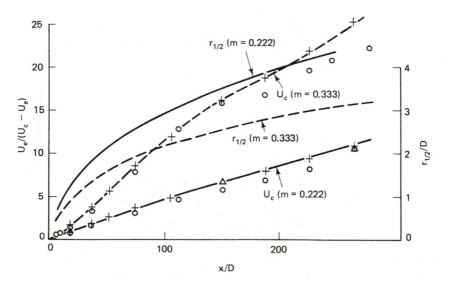

Figure 8-2 Variation of centerline velocity and half-radius for an axisymmetric jet: $\bigcirc$, $\triangle$, hot wire; $+$ Pitot tube. (From Antonia and Bilger, 1973.)

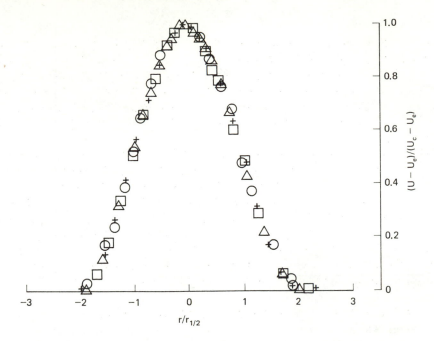

Figure 8-3 Nondimensional radial profiles across an axisymmetric jet: $m = 0.222$; $x/D = 38$ (+), 76 ($\triangle$), 152 ($\square$), 266 ($\bigcirc$). (From Antonia and Bilger, 1973.)

fraction of the mean flow velocity difference. The streamwise behavior of the centerline values and the characteristic shape of the transverse profiles can be easily seen. The maximum generally occurs somewhat off the centerline, since the mean velocity gradient and hence the turbulence production is greater there [see Fig. 8-3 and Eq. (7-89)]. Observe that the profiles of this variable

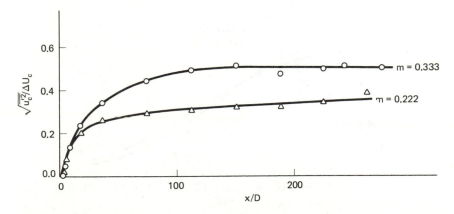

Figure 8-4 Streamwise variation of centerline axial turbulence intensity in axisymmetric jets. (From Antonia and Bilger, 1973.)

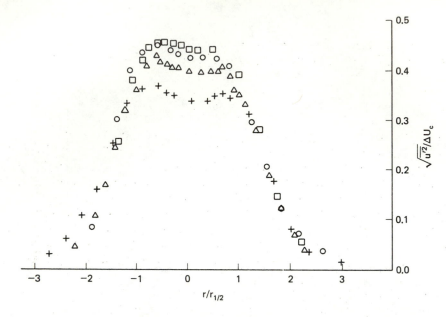

Figure 8-5 Radial profiles of axial turbulence intensity in axisymmetric jets, $x/D = 38$ (+), 76 ($\triangle$), 152 ($\square$), 266 ($\bigcirc$); $m = 0.333$. (From Antonia and Bilger, 1973.)

have not attained a similarity condition by $x/D \approx 200$. The data of Gibson (1963) for tests with $U_e = 0$ showed that the relative magnitudes of the axial, transverse, and peripheral turbulence intensities were about the same.

The variation of the turbulent kinetic energy across the layer, again for $U_e = 0$, is shown in Fig. 8-6 from Rodi (1972) and Wygnanski and Fiedler

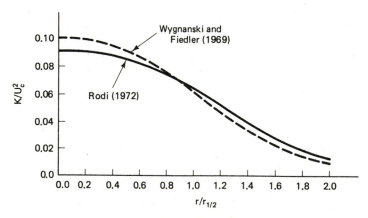

Figure 8-6 Radial profile of turbulent kinetic energy for an axisymmetric jet with $m = 0$.

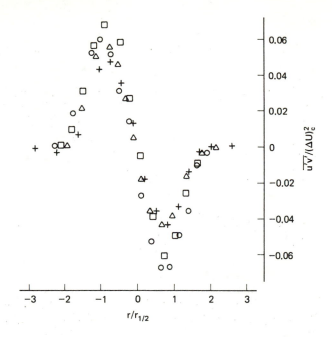

Figure 8-7 Radial profiles of Reynolds stress for an axisymmetric jet with $m = 0.333$; $x/D = 38$ (+), 96 ($\triangle$), 152 ($\square$), 248 ($\bigcirc$). (From Antonia and Bilger, 1973.)

(1969). Information on the Reynolds turbulent shear stress is available for $U_e \neq 0$, and this is shown in Fig. 8-7. Note the antisymmetry of this quantity in a symmetrical flow. About a line (or plane) of symmetry, $\overline{u'v'} = 0$ when $\partial U/\partial y = 0$.

Spectra of the axial turbulence intensity are shown in Fig. 8-8. Again, motion of several decades in wave number k_1 can be observed. Finally, we present the energy balance across the flow where the magnitude of the individual terms in Eq. (7-89) can be seen and studied. The most complete and reliable information available is the basic data of Wygnanski and Fiedler (1969) for $U_e = 0$ as modified by Rodi (1975). This balance is presented in Fig. 8-9. The magnitude of the various terms across the layer is much different than for a boundary layer (compare with Fig. 7-24). In this case, the energy production by the normal stresses in addition to production by shear stress,

$$-(\overline{u'^2} - \overline{v'^2})\frac{\partial U}{\partial x} \tag{8-3}$$

was included for completeness. It can be seen that this contribution is the smaller of the two, except near the axis.

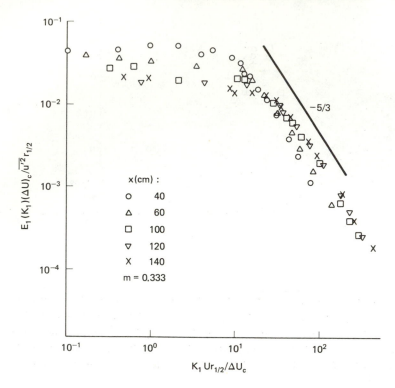

Figure 8-8 Axial-component spectra for an axisymmetric jet. (From Antonia and Bilger, 1973.)

8-2-2 Density Variations from Temperature Variations

The simplest cases with density variations are produced with fluid injection at a temperature different than the main flow. Experiments of that type in the axisymmetric geometry were reported by Landis and Shapiro (1951). Mean flow results in terms of the axial variation of the centerline velocity and temperature are shown here in Fig. 8-10 for a case with $T_e/T_j = 0.77$ and $U_e/U_j \equiv m = 0.50$. (The density ratio $\rho_e/\rho_j \equiv n$ for such cases is related to the temperature ratio as $T_e/T_j = \rho_j/\rho_e = 1/n$.) It can be seen that the power decay law exponent P_i (i.e., $\Delta \bar{T} \sim x^{P_T}$, $\Delta U \sim x^{P_V}$, etc.) is slightly greater for this case than that for the nearly constant density case presented by Forstall and Shapiro (1950).

The axial variations of the half-widths for the velocity and temperature profiles are shown in Fig. 8-11 for the same case as in Fig. 8-10. From these data, we can see the important result that nondimensionalized temperature profiles are always wider or *fuller* than the corresponding velocity profiles, which indicates that the transverse transport of thermal energy by turbulence is more rapid than that for momentum. This means that a turbulent thermal

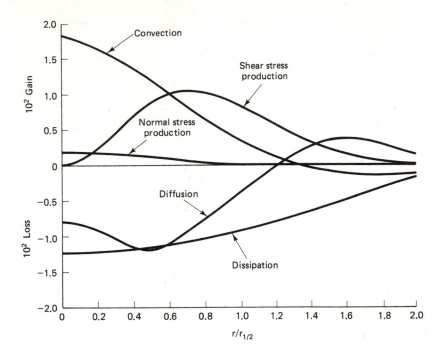

Figure 8-9   Turbulent energy balance for an axisymmetric jet with $m = 0$. All terms $\times \; r_{1/2}/U_c^3$. (From Wygnanski and Fiedler, 1969, as modified by Rodi, 1975.)

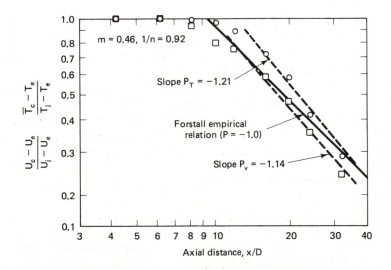

Figure 8-10   Streamwise variation of centerline velocity and temperature in a heated, axisymmetric jet. (From Landis and Shapiro, 1951.)

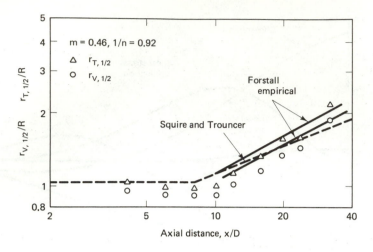

Figure 8-11   Variation of half-radii based on velocity and temperature in a heated, axisymmetric jet. (From Landis and Shapiro, 1951.)

conductivity k_T defined by

$$-k_T \frac{\partial \bar{T}}{\partial y} = \rho c_p \overline{v' T'} = q_T \tag{8-4}$$

when combined with $v_T \equiv \mu_T/\rho$, to give a *turbulent* Prandtl number

$$\mathrm{Pr}_T = \frac{v_T}{k_T/\rho c_p} \tag{8-5}$$

will correspond to values of $\mathrm{Pr}_T < 1$. For the axisymmetric case of Corrsin and Uberoi (1949), this quantity was determined to be approximately 0.7. Planar jet data indicate a value of $\mathrm{Pr}_T \approx 0.5$.

For turbulence data on heated jets, the recent work of Chevray and Tutu (1978) with $U_e = 0$ has been chosen. Velocity and temperature fluctuations at $x/D = 15$ are given in Fig. 8-12. The radial variations of turbulent shear and heat transfer are shown in Fig. 8-13. These profiles have similar shapes, and their peaks occur at roughly the same radial station, which is not, however, at the location of the maximum value of either $\partial U/\partial y$ or $\partial \bar{T}/\partial y$. These facts suggest that the transport mechanisms for momentum and heat are quite similar and that gradient transport models are suspect even for this simple flow problem.

Finally, high-speed, free-turbulent-flow situations such as the wake behind a supersonic body produce substantial temperature variations. There has been no evidence to indicate, however, that significant new processes occur beyond those that occur in heated, low-speed, jets, at least as far as the mean flow is concerned.

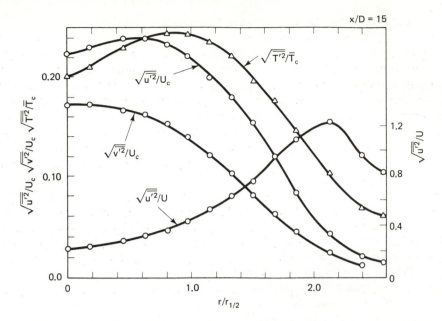

Figure 8-12   Radial profiles of velocity and temperature fluctuations for an axisymmetric jet with $m = 0$. (From Chevray and Tutu, 1978.)

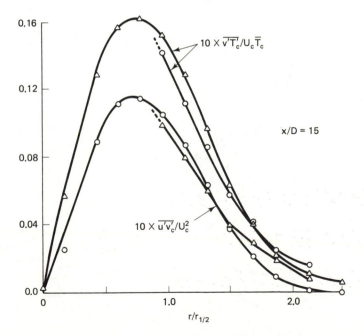

Figure 8-13   Radial profiles of turbulent shear and heat flux for an axisymmetric jet with $m = 0$. (From Chevray and Tutu, 1978.)

8-2-3 Density Variations from Composition Variations

Many important practical applications involve injection of one fluid into surroundings of a different fluid. In this book, the discussion is restricted to cases of one phase, either gas or liquid, and binary mixtures.

An experimental correlation for the length of the *potential core* in terms of concentration of injected species was developed by Chriss (1968) as

$$\frac{x_c}{D} = 6.5 \sqrt{\frac{\rho_j U_j}{\rho_e U_e}} \tag{8-6}$$

For data on the influence of composition variations on the axial decay of centerline flow variables, the tests of Chriss (1968) can be used again. In Fig. 8-14, we have the influence of n and m on the centerline velocity decay. The influence is obviously strong. The axial variation of the centerline composition of injectant is presented in Fig. 8-15.

When the axial decay of centerline values is plotted on logarithmic paper, a power-law decay rate is again shown. Many workers have attempted to determine the values of the relevant exponents for each type of variable: velocity, temperature, and composition. Schetz (1969) looked at a large group

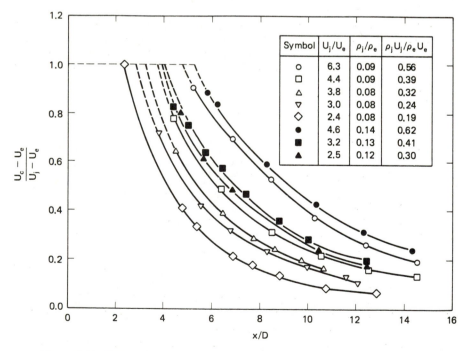

Figure 8-14 Streamwise variation of centerline velocity for axisymmetric H_2 jets into air. (From Chriss, 1968.)

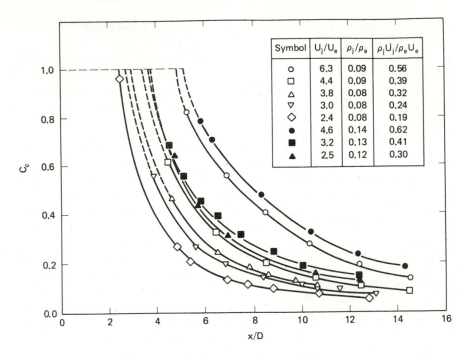

Figure 8-15 Streamwise decay of centerline concentration for an axisymmetric H_2 jet into air. (From Chriss, 1968.)

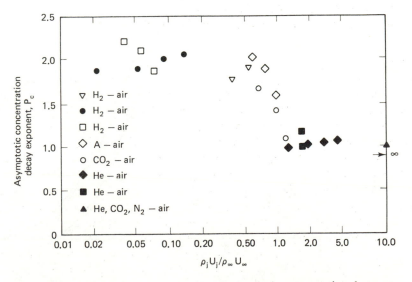

Figure 8-16 Experimental determination of asymptotic concentration decay exponent. (From Schetz, 1969.)

of data from various investigators to develop a correlation for P_c in terms of the single parameter $\rho_j U_j / \rho_e U_e$. The result is shown here in Fig. 8-16.

As for density changes produced by temperature variations alone, the half-width of concentration profiles is larger than for velocity, and the profiles themselves are consequently *fuller*, as shown in Fig. 8-17. Indeed, nondimensional temperature and concentration profiles have been found to be virtually identical.

Introducing a turbulent mass diffusion coefficient D_T through

$$- \rho D_T \frac{\partial C_i}{\partial y} = \rho \overline{v' c_i'} \qquad (8\text{-}7)$$

we may express this quantity in relation to turbulent momentum or heat transfer through a turbulent Schmidt number

$$\mathrm{Sc}_T = \frac{\nu_T}{D_T} \qquad (8\text{-}8)$$

or a turbulent Lewis number

$$\mathrm{Le}_T = \frac{D_T}{k_T / \rho c_p} = \frac{\mathrm{Pr}_T}{\mathrm{Sc}_T} \qquad (8\text{-}9)$$

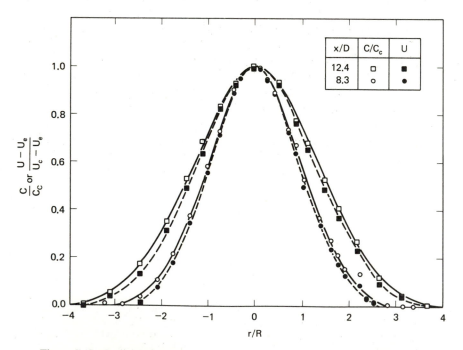

Figure 8-17 Radial profiles of velocity and concentration for an axisymmetric H_2 jet into air. (From Chriss, 1968.)

Most workers agree that for the axisymmetric case $Pr_T \approx Sc_T \approx 0.7$, which leads to $Le_T \approx 1.0$. For planar flows, $Pr_T \approx Sc_T \approx 0.5$ and $Le_T \approx 1.0$.

The accurate measurement of quantities involving concentration fluctuations is generally conceded to be a difficult task, and various methods have been tried. The experiments of Becker et al. (1967) were for the axisymmetric geometry with $U_e = 0$, and the variable composition was provided by oil smoke mixed with the air in the injectant. The measurements were made using the light-scattering technique. Raman scattering of laser light was used to obtain detailed mean and turbulent measurements in a round natural gas (95% methane) jet exhausting into air by Birch et al. (1978). Radial profiles of the root mean square of the normalized concentration fluctuations from both groups at various x/D are shown in Fig. 8-18. These profiles are in close agreement with the corresponding measurements for temperature by Antonia

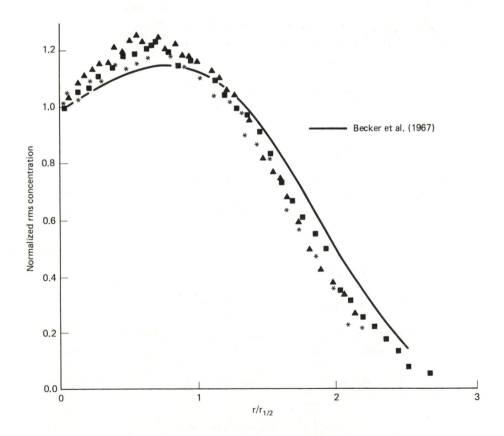

Figure 8-18 Radial profiles of concentration fluctuation for an axisymmetric methane jet into stationary air $x/D = 20$ (■), 30 (▲), 40 (∗). (From Birch et al., 1978.)

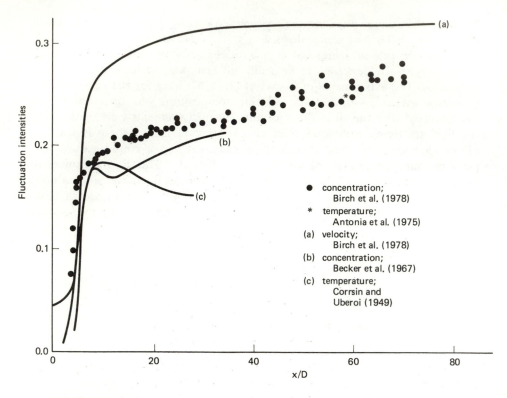

Figure 8-19 Streamwise variation of centerline fluctuation intensities for an axisymmetric jet with $m = 0$.

et al. (1975). The axial variation of the intensity of the concentration fluctuations is shown in Fig. 8-19, together with corresponding results for the velocity and temperature fields.

8-3 MEAN FLOW ANALYSES

8-3-1 Similarity Analyses

The oldest and simplest analyses that exist are based on the observed similarity of profiles, as was displayed in Figs. 8-3 and 8-17. Taking the case of a constant-density round jet with $U_e = 0$ as an extremely easy example, one need only assert that the integral of the streamwise momentum remains constant at its initial value

$$ J = \rho \int U^2 \, dA = \text{constant} = \rho_j U_j^2 A_j \qquad (8\text{-}10) $$

and that the width of the mixing zone $b(x) \sim x$ from experiment to proceed.

Since the profile may be taken as *similar* for $x/D \gg 1$, we can write

$$\frac{U(x, r)}{U(x, 0)} = f\left(\frac{r}{b}\right) \tag{8-11}$$

Substituting into Eq. (8-10) one gets

$$U(x, 0) = \text{constant} \left(\frac{1}{b}\right)\sqrt{\frac{J}{\rho}} \tag{8-12}$$

where the *constant* depends on the particular *shape* of the profile [i.e., the form of $f(r/b)$]. Finally,

$$U_c(x) = U(x, 0) = \text{constant} \left(\frac{1}{x}\right)\sqrt{\frac{J}{\rho}} \tag{8-12a}$$

since b is proportional to x. The corresponding result for a planar jet with $U_e = 0$ is $U_c \sim 1/\sqrt{x}$. The analyses can and have been carried further to produce complete solutions for the velocity profiles within an unknown constant by Tollmien (1926) and Görtler (1942). The unknown constant must be determined by experiment, and then good agreement with the shape of experimental profiles is obtained as shown in Fig. 8-20. These solutions are not, however, easily extended to cases with $U_e \neq 0$.

For the wake behind a body, the profiles only become *similar* at a distance of many characteristic body dimensions downstream where the velocity defect $\Delta U \equiv (U_\infty - U)$ has become small compared to U_∞. Also, at such a location, the static pressure will return to the freestream value ahead of the body. The *momentum theorem* can then be applied to a control volume that contains the body and whose downstream surface is in the *similar* profile

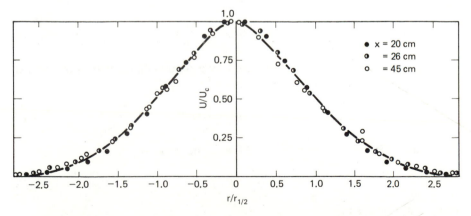

Figure 8-20 Velocity distribution in a circular, turbulent jet. (Measurements due to Reichardt, 1942; Analysis due to Tollmien, 1926.)

region. If the body is a general, two-dimensional (planar) cylinder, we may write for a body length L perpendicular to the plane of the flow:

$$\text{drag} = \rho L \int_{-\infty}^{\infty} U(U_\infty - U) \, dy \qquad (8\text{-}13)$$

Noting, the definition of ΔU and the fact $U \approx U_\infty$ (since $\Delta U \ll U_\infty$), Eq. (8-13) can be approximated as

$$\text{drag} \approx \rho U_\infty L \int_{-\infty}^{\infty} (\Delta U) \, dy \qquad (8\text{-}13\text{a})$$

The value of the integral will be proportional to $\Delta U_c \, b(x)$ with the proportionality factor depending on the details of the profile shape $\Delta U(y)$. Here ΔU_c is the centerline value of ΔU. With this and the definition of a drag coefficient as $C_D = \text{drag}/(\tfrac{1}{2}\rho U_\infty^2 \, LD)$, where D is a characteristic thickness of the cylinder (e.g., the diameter for a circular cylinder), we can set drag as expressed by the two relations equal and obtain

$$\frac{\Delta U_c}{U_\infty} \sim \frac{C_D D}{2b} \qquad (8\text{-}14)$$

The growth of the width $b(x)$ is determined based on the assumption that the time rate of increase of the width is proportional to the transverse velocity fluctuation v'; that is,

$$\frac{Db}{Dt} \sim v' \qquad (8\text{-}15)$$

with

$$\frac{Db}{Dt} \approx U_\infty \frac{db}{dx} \qquad (8\text{-}15\text{a})$$

and

$$v' \sim \ell_m \frac{\partial U}{\partial y} \approx \ell_m \frac{\Delta U_c}{b} \qquad (8\text{-}16)$$

Taking, as usual for a flow not restricted by a rigid wall, ℓ_m as proportional to the width of the layer (i.e., $\ell_m = K_2 b$), one obtains

$$\frac{db}{dx} \sim K_2 \frac{\Delta U_c}{U_\infty} \qquad (8\text{-}17)$$

Now, substitute for ΔU_c from Eq. (8-14) and the result is

$$2b \frac{db}{dx} \sim K_2 C_D D \qquad (8\text{-}18)$$

Integration results in

$$b \sim (K_2 \, x C_D \, D)^{1/2} \qquad (8\text{-}19)$$

Using this in Eq. (8-14) yields

$$\frac{\Delta U_c}{U_\infty} \sim \left(\frac{C_D D}{K_2 x}\right)^{1/2} \tag{8-20}$$

Comparison with experiment gives the values of the proportionality constants and leads to

$$b = 0.57(x C_D D)^{1/2} \tag{8-19a}$$

$$\frac{\Delta U_c}{U_\infty} = 0.98\left(\frac{C_D D}{x}\right)^{1/2} \tag{8-20a}$$

The corresponding results for a circular wake are

$$b \sim (K_2 C_D A x)^{1/3} \tag{8-21}$$

and

$$\frac{\Delta U_c}{U_\infty} \sim \left(\frac{C_D A}{K_2^2 x^2}\right)^{1/3} \tag{8-22}$$

8-3-2 Analyses Based on Eddy Viscosity or Mixing-Length Models

Over the years, various approximate methods for treating jet mixing problems with $U_e \neq 0$ and the far wake have been presented. These methods were generally concerned primarily with overcoming the difficulties of solving the equations of motion. The widespread availability of large digital computers that permit *numerically exact* solutions has rendered these methods largely obsolete, and they are not covered here.

Since it has now been assumed that the equations of motion can be solved (e.g., numerically), within the limits of mean flow turbulence models the discussion reduces to a choice of an eddy viscosity or mixing-length model. The major models that have been used are listed here in Table 8-1. Prandtl's third model (which is the most widely used of the first five for jets) introduced a new concept where $v_T \sim |\Delta U|$. Note however that $\delta^* \sim \Delta U$, so the Clauser model contains that dependence also. It is important to observe that the three entries following Prandtl's third model can be shown to be equivalent to it and to each other. Consider first the *extended* Clauser model,

$$v_T = \frac{\mu_T}{\rho} = C U_e \delta^* \tag{8-23}$$

where $C \approx 0.018$ and δ^* must be interpreted to be based on the absolute value of $[1 - (U/U_e)]$. Now compare this to the Prandtl model, using the half-width $b_{1/2}(x)$:

$$v_T = 0.037 b_{1/2} |U_{max} - U_{min}| \tag{8-24}$$

TABLE 8-1 Kinematic Eddy Viscosity Models for Main Mixing Region of Jets and Wakes

Author	Planar	Axisymmetric	Variable Density	Expression	Remarks
Prandtl (1925)	×	×		$\ell_m^2\left(\dfrac{\partial U}{\partial y}\right)$	ℓ_m proportional to the width of the mixing region
von Kármán (1930)	×	×		$\kappa^2\dfrac{(\partial U/\partial y)^3}{(\partial^2 U/\partial y^2)^2}$	
Taylor (1932)	×	×		$\tfrac{1}{2}\ell_w^2\left(\dfrac{\partial U}{\partial y}\right)$	$\ell_w = \sqrt{2}\,\ell_m$
Prandtl (1942)	×	×		$\ell_m\sqrt{\left(\dfrac{\partial U}{\partial y}\right)^2 + \ell_1^2\left(\dfrac{\partial^2 U}{\partial y^2}\right)^2}$	Requires two mixing lengths
Prandtl (1942)	×	×		$\kappa_1 b(U_{max} - U_{min})$	Introduced "velocity difference" concept; with b taken as $b_{1/2}$, $\kappa_1 = 0.037$ in planar jets; Schlichting extended this to axisymmetric jets with $r_{1/2}$ and $\kappa_1 = 0.025$
Schlichting (1942)	×			$0.0222 U_e C_D D$	Wake of a cylinder of arbitrary cross section
Clauser (1956)				$CU_e\delta^* = C\displaystyle\int_0^\infty |V_e - U|\,dy$	Applied to "wake"-like outer region of a boundary layer $0.016 < C < 0.018$
Hinze (1959)	×			$0.016 U_e D$	Wake of a Circular Cylinder
Ting-Libby (1960)		×	×	$\rho^2\nu_T = \dfrac{\rho_c^2\nu_{T_0}}{r^2}\displaystyle\int_0^r 2\frac{\rho}{\rho_c}r'\,dr'$	ν_{T_0} is the constant density eddy viscosity and ρ_c is the centerline density
Ting-Libby (1960)	×		×	$\rho^2\nu_T = \rho_c^2\nu_{T_0}$	
Ferri et al. (1962)		×	×	$\rho\nu_T = 0.025 r_{1/2}[(\rho U)_{max} - (\rho U)_{min}]$	Extended Prandtl's third model to variable density introduced "mass flow difference" concept
Schetz (1964)	×		×	$\rho^2\nu_T = 0.037 b_{1/2}\rho_c[(\rho U)_{max} - (\rho U)_{min}]$	Application of "mass flow difference" to planar flows
Schetz (1968)	×		×	"Turbulent viscosity proportional to mass flow defect (or excess) in the mixing region"	

For simple profile shapes such as a rectangular, triangular, or cosine velocity defect (or excess), one finds that the two expressions agree exactly in form and to the extent of 0.036 versus 0.037 as the proportionality constant. The wake models of Schlichting and Hinze can be reduced to the same form as the extended Clauser model. Noting that

$$C_D D = 2\Theta\left(\equiv 2 \int_{-\infty}^{\infty} \frac{U}{U_e}\left(1 - \frac{U}{U_e}\right) dy\right) \qquad (8\text{-}25)$$

and taking a crude representative value of $C_D \approx 1.20$ for a circular cylinder, these expressions become, respectively,

$$v_T = 0.044 U_e \Theta \qquad (8\text{-}26a)$$

$$v_T = 0.027 U_e \Theta \qquad (8\text{-}26b)$$

In the treatment of wake problems, it is common to neglect the factor U/U_∞ in the definition of Θ, since $U/U_\infty \approx 1$ (see the preceding section). This renders

$$\Theta \approx \Delta^*\left(\equiv \int_{-\infty}^{\infty} \left|1 - \frac{U}{U_e}\right| dy \right) \qquad (8\text{-}27)$$

so that these formulas could as well be written as

$$v_T = 0.044 U_e \Delta^* \qquad (8\text{-}28a)$$

$$v_T = 0.027 U_e \Delta^* \qquad (8\text{-}28b)$$

The extended Clauser model written in these terms is

$$v_T = 0.018 U_e \Delta^* \qquad (8\text{-}29)$$

where we have taken the displacement thickness appropriate to a *two-sided,* planar free mixing problem rather than the *one-sided* boundary layer case considered by Clauser. It will be shown below that Eq. (8-29) provides predictions in good agreement with jet experiments. The question arises as to why the constant for wakes ($U_c/U_e < 1$) is so much larger than that for jets ($U_c/U_e > 1$). Abramovich (1960) notes this effect and attributes it to increased turbulence caused by the separated base flow in the wake case. For our purposes here, however, the important result is that these free mixing eddy viscosity models are all equivalent in functional form.

The models listed in Table 8-1 following the wake models are attempts to extend the basic Prandtl model to problems involving significant density variations. The Ting–Libby model results from an attempt to apply transformation theory to turbulent free mixing; it has been shown to be unreliable. Ferri's suggestion of utilizing a mass flow difference to replace the velocity difference in the Prandtl model has provided predictions of unreliable accuracy for the axisymmetric case. However, when the mass flow difference was applied to the planar case, a good prediction was achieved (Schetz, 1964).

Since it was possible to demonstrate some unity of the models for planar, constant-density cases, it was instructive to examine new means for extending these models to the compressible case. As these models are all equivalent, one could begin with any one. Rather than the usual procedure of starting with the Prandtl model, Schetz (1968) chose to begin with the extended Clauser model. It was simple, at least formally, to extend this further to varying density, that is,

$$\nu_T = 0.018 U_e \Delta^* = 0.018 U_e \int_{-\infty}^{\infty} \left| 1 - \frac{\bar{\rho} U}{\rho_e U_e} \right| dy \qquad (8\text{-}30)$$

and to show that for simple profile shapes this expression is equivalent to the planar mass flow difference model given in Table 8-1. Recalling that the model produced predictions in good agreement with experiment, it may be stated that the extended Clauser model, Eq. (8-30), can be viewed as an adequate representation of planar, free mixing flows with or without strong density variations. To support that assertion, we present in Figs. 8-21 and 8-22 some comparisons of experimental data and predictions based on Eq. (8-30) from Schetz (1971). It should be reemphasized here that the Prandtl model and the extended Clauser model are essentially equivalent in the planar case, so that the same level of agreement could be obtained with the Prandtl model.

This still left the axisymmetric case without a satisfactory eddy viscosity model, especially for variable density cases. Schetz (1968) generalized the extended Clauser model further to the axisymmetric geometry as follows. Rewrite Eq. (8-30) as

$$\mu_T = \bar{\rho} \nu_T = 0.018 \bar{\rho} U_e \Delta^* \qquad (8\text{-}30a)$$

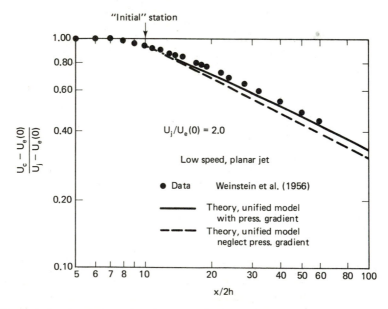

Figure 8-21 Comparison of prediction and experiment for centerline velocity variation of a planar jet. (From Schetz, 1971.)

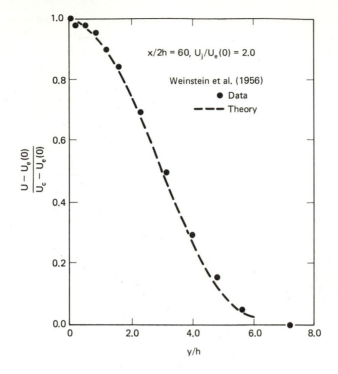

Figure 8-22 Comparison of prediction and experiment for transverse velocity profile in a planar jet. (From Schetz, 1971.)

This can be read to say: "The turbulent viscosity is proportional to the mass flow defect (or excess) per unit width in the mixing region." One can carry this statement over into axisymmetric flow by defining a new displacement thickness δ_r^* as

$$\pi \rho_e U_e (\delta_r^*)^2 \equiv \int_0^\infty |\rho_e U_e - \bar{\rho} U| \, 2\pi r \, dr \qquad (8\text{-}31)$$

This follows directly from the logic used for the ordinary displacement thickness (see Sec. 2-2) and noting the geometry. Using Eq. (8-31) in the statement quoted above gives

$$\mu_T = \frac{C_3 \rho_e U_e \pi (\delta_r^*)^2}{R} \qquad (8\text{-}32)$$

The proportionality constant C_3 had to be determined by a comparison between theory and experiment for one case, as is done with all eddy viscosity models, and the experiments of Forstall and Shapiro (1950) were employed. The results are shown in Fig. 8-23; the value of $C_3 \pi$ determined is 0.018. This constant is held fixed at that value for all further calculations. One can observe

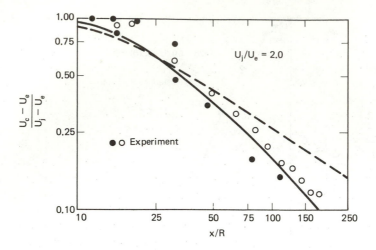

Figure 8-23 Comparison of prediction and experiment for centerline velocity variation of an axisymmetric jet. Dashed line: analysis with Schlichting axisymmetric version of Prandtl model; Solid line: analysis with Eq. (8-32). (Data from Forstall and Shapiro, 1950. From Schetz, 1968.)

that this model gives excellent qualitative as well as quantitative agreement with the data. Schlichting's extension of the Prandtl model to axisymmetric jets produces poor predictions for the centerline velocity decay.

To consider variable density cases for a jet or wake, it is necessary to add energy and/or species conservation equations to the system. Heat transfer is, in general, simply modeled using an eddy viscosity or mixing-length model and a constant value of the *turbulent* Prandtl number [see Eqs. (8-4) and (8-5)]. Mass transfer is correspondingly modeled using a constant value of the *turbulent* Schmidt or Lewis number [see Eqs. (8-7) to (8-9)]. The case of a jet of hydrogen injected into an airstream provides a stringent test of the theory, since there is a very large density gradient across the mixing zone. The data of Chriss (1968) were used to make the comparisons of theory and experiment, and the Ferri model and Eq. (8-32) were used with a constant *turbulent* Schmidt number of 0.75 (Schetz, 1968). Results for the particular case with $\rho_i U_j / \rho_\infty U_\infty = 0.56$, $U_j / U_\infty = 6.3$, are shown in Fig. 8-24. The calculation was started with experimental profiles at $x/R = 5.9$. Here, as has been generally true, the model of Eq. (8-32) produced a superior prediction. In addition to the prediction of the behavior of centerline values, the adequacy of the prediction of transverse profiles is also of interest. A comparison of prediction and experiment is given as Fig. 8-25. Again, good agreement is obtained using the eddy viscosity model of Eq. (8-32).

Aside from the philosophical objection that the mean flow models (eddy viscosity or mixing-length) do not directly reflect the actual turbulent nature of the flow, real cases occur where that deficiency becomes of physical impor-

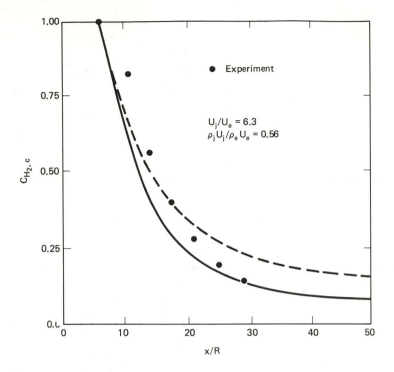

Figure 8-24 Comparison of prediction and experiment for centerline concentration decay for an axisymmetric H_2 jet into air. Dashed line: analysis with Ferri Model; Solid line: analysis with Eq. (8-32). (Data from Chriss, 1968. From Schetz, 1968.)

tance. One such instance has been alluded to before. Eddy viscosity models that have been successfully applied to jet or wake cases often do not perform adequately when applied to the other cases. This is so even when the functional forms of both kinds of models are essentially identical. The discrepancy is primarily centered on the value of the proportionality constant, and the difference is not simply academic. We have shown that Eq. (8-29) is capable of good predictions of planar jet mixing flows. What happens if one uses it for a wake case? Figure 8-26 shows such a comparison for the wake behind a circular cylinder using the data of Schlichting (1930). Good agreement with experiment can be achieved only with an increase in the value of the proportionality constant significantly above that successfully used for jet problems.

In Schetz (1971), the possible direct connection between the observed turbulence in various flow problems and the proportionality constant in eddy viscosity models was investigated. Using the planar, constant-density case as an example, one starts with the definition of the eddy viscosity

$$-\rho \overline{u'v'} = \mu_T \frac{\partial U}{\partial y} = \rho \nu_T \frac{\partial U}{\partial y} \qquad (7\text{-}36)$$

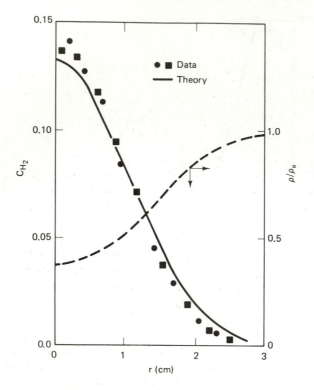

Figure 8-25 Comparison of prediction and experiment for a radial concentration profile in an axisymmetric H_2 jet into air. (Data from Chriss, 1968; analysis with Eq. (8-32) from Schetz, 1968.)

and approximates the velocity gradient crudely as $\partial U/\partial y \approx \Delta U_c/b$. Working with the extended Clauser model and replacing the numerical value of the now unknown proportionality constant by C_4, we have, using the preceding approximation,

$$-\rho\overline{u'v'} = C_4\rho \int_{-\infty}^{\infty} |U_e - U|\,dy\left(\frac{\Delta U_c}{b}\right) \qquad (8\text{-}33)$$

The displacement thickness integral can be represented by a constant C_5 times the product of the velocity difference ΔU_c and the width b. The actual value of C_5 depends on the specific profile shape, but typical shapes yield a value of roughly one-half. Thus,

$$-\overline{u'v'} \approx C_4 C_5 (\Delta U_c)^2 \qquad (8\text{-}34)$$

In general, one may write

$$-\overline{u'v'} \approx C_6 \sqrt{\overline{u'^2}}\sqrt{\overline{v'^2}} \qquad (8\text{-}35)$$

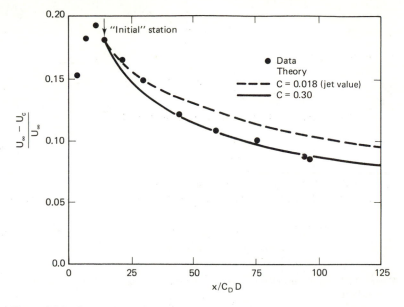

Figure 8-26 Comparison of prediction and experiment for centerline velocity in the wake behind a circular cylinder. (Data from Schlichting, 1930; analysis with Eq. (8-29) using different constants from Schetz, 1971.)

where C_6 is also about one-half. Further, in shear flows

$$u' \approx -v' \tag{8-36}$$

so that the final result emerges as

$$\frac{\overline{(u')^2}}{(\Delta U_c)^2} \approx C_4\left(\frac{C_5}{C_6}\right) \tag{8-37}$$

where C_5/C_6 is expected to be near unity. Equation (8-37) achieves the desired result of relating the proportionality constant to a characteristic of the turbulence. Some data for the turbulence/mean flow information that appears in Eq. (8-37) are presented in Fig. 8-27. The fact that this quantity is more or less constant for a given flow problem, except in the near-wake or *potential core*, is comforting in light of Eqs. (8-29) and (8-37), since C_4 is presumed to be a constant. Comparing the results for the planar jet with $U_e \neq 0$ with those for the wake behind a circular cylinder, we see that the increase of C_4 by a factor of $\frac{5}{3}$ required in going from a jet to a wake behind a circular cylinder as found for the case in Fig. 8-26, is roughly predicted by Eq. (8-37). For wakes behind streamlined (unseparated) bodies, $\overline{u_c'^2}/(\Delta U_c)^2$ is quite close to the values found in jets, so the appropriate proportionality constant is correspondingly the same as for jets.

With this, the first-order influence of turbulence structure can be crudely

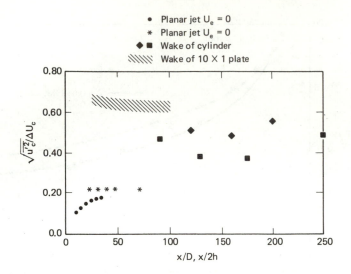

Figure 8-27 Axial turbulence intensities for planar mixing flows collected by Schetz, 1971.

incorporated into an eddy viscosity (or mixing-length) model, as long as information of the type in Fig. 8-27 is available for the general flow problem under consideration.

8-4 ANALYSES BASED ON TURBULENT KINETIC ENERGY

The development of the two basic TKE models in use for wall-bounded flows was described in Sec. 7-10. For free mixing flows, they are used in the same forms except that it is not necessary to modify ℓ for wall effects.

The very active group at Imperial College contributed a comprehensive survey paper (Launder et al., 1971) to the 1970 NASA Conference on Free Turbulent Shear Flows which compared the adequacy of several turbulence models with free mixing data, all on the same basis. At the TKE level, they used the Prandtl energy method [Eq. (7-91)]. Harsha (1971a) extended the Bradshaw TKE approach [Eq. (7-92)] to free shear flows.

We begin a discussion of some comparisons of predictions with experiment with the case with $U_j/U_\infty = 4.0$ from Forstall and Shapiro (1950). In Fig. 8-28 the centerline velocity predictions based on the eddy viscosity model, Eq. (8-32) from Schetz (1971), the mixing-length model, the TKE model based on Eq. (7-91) from Launder et al., (1971), and the TKE model based on Eq. (7-92) from Harsha (1971a) are shown, compared with experiment. The predictions are all in reasonable agreement with the data. It appears that the various

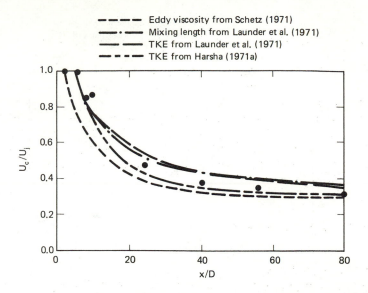

Figure 8-28 Comparison of predictions with several models with the axisymmetric jet experiments of Forstall and Shapiro, 1950.

models would all give predictions even closer together if they were started at the end of the potential core ($x/D \approx 7$) rather than at the station specified by the NASA conference organizers, which was still in the potential core.

The H_2–air mixing cases of Chriss (1968) provide test cases with very

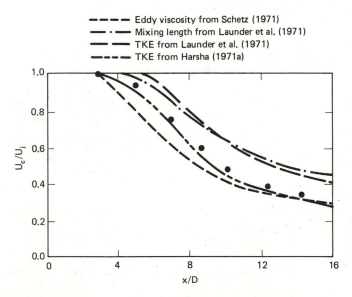

Figure 8-29 Comparison of predictions with several models with the axisymmetric H_2 jet into air experiment of Chriss, 1968: centerline velocity.

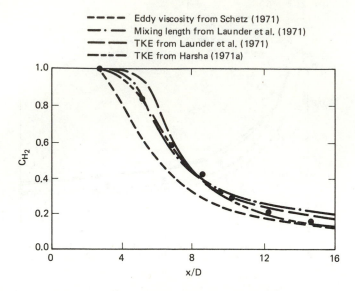

Figure 8-30 Comparison of predictions with several models with the axisymmetric H_2 jet into air experiment of Chriss, 1968: centerline concentration.

large density differences across the mixing layer. Figures 8-29 and 8-30 show the centerline variations of velocity and concentration, comparing the same models as for Fig. 8-28 with the data. Again in this case it is hard to detect any clear advantage of either TKE model over the mean flow models.

8-5 ANALYSES BASED ON EQUATIONS FOR TURBULENT KINETIC ENERGY AND LENGTH SCALE

The basic material required here was developed in Sec. 7-11, but again it is not necessary in the present case to treat a special wall-dominated region. Launder et al. (1971) used $Z \equiv \varepsilon$, and the final modeled equation for ε was taken as (note: $j \equiv 0$ for planar flows and $j \equiv 1$ for axisymmetric flows)

$$\rho\left(U\,\frac{\partial \varepsilon}{\partial x} + V\,\frac{\partial \varepsilon}{\partial y}\right)$$

$$= \frac{1}{y^j}\,\frac{\partial}{\partial y}\left(y^j\,\frac{\mu_T}{\sigma_\varepsilon}\,\frac{\partial \varepsilon}{\partial y}\right) + C_{\varepsilon 1}\,\frac{\rho \varepsilon}{K}\,\nu_T\left(\frac{\partial U}{\partial y}\right)^2 - C_{\varepsilon 2}\,\frac{\rho \varepsilon^2}{K} \qquad (8\text{-}38)$$

with $\sigma_\varepsilon = 1.3$, $C_{\varepsilon 1} = 1.43$, and $C_{\varepsilon 2} = 1.92$ for planar flows. For axisymmetric flows,

$$C_{\varepsilon 2} = 1.92 - 0.0667F \qquad (8\text{-}39)$$

where

$$F = \frac{b}{2\Delta U_c} \left(\frac{dU_c}{dx} - \left| \frac{dU_c}{dx} \right| \right) \tag{8-40}$$

The TKE equation is retained with $\sigma_K = 1.0$ and

$$\nu_T = \frac{C_\mu K^2}{\varepsilon} = \frac{(0.09 - 0.04F)K^2}{\varepsilon} \tag{8-41}$$

This model is called $K\varepsilon 1$. For weak shear flows (i.e., for cases where the velocity defect or excess ΔU_c is a small fraction of U_e), an extended version called $K\varepsilon 2$ was presented. This model used the same equations, but now with $C_{\varepsilon 1} = 1.40$, $C_{\varepsilon 2} = 1.94$, $\sigma_\varepsilon = 1.0$, and

$$C_\mu = 0.09g\left(\frac{\bar{\mathscr{P}}}{\varepsilon}\right) \tag{8-42}$$

where

$$\frac{\bar{\mathscr{P}}}{\varepsilon} = \int_0^b \tau_T \left(\frac{\mathscr{P}}{\varepsilon}\right) y^j \, dy \bigg/ \int_0^b \tau_T y^j \, dy \tag{8-43}$$

$\mathscr{P}$ is production of K, and $g(\bar{\mathscr{P}})/\varepsilon)$ is given as a graph.

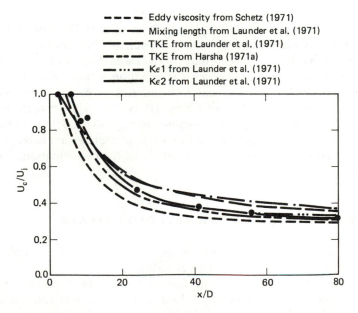

Figure 8-31 Comparison of predictions with mean flow, one-equation, and two-equation models with the axisymmetric jet experiment of Forstall and Shapiro, 1950.

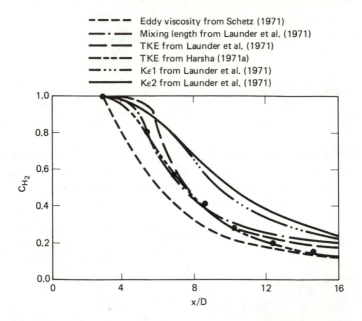

Figure 8-32 Comparison of predictions with mean flow, one-equation, and two-equation models with the axisymmetric H_2 jet into air experiment of Chriss, 1968.

Consider the low-speed, axisymmetric jet case of Forstall and Shapiro (1950) which was discussed before; comparison of predictions and data are given in Fig. 8-31. It can be seen that the $K\varepsilon 1$ and $K\varepsilon 2$ models perform very well. For axisymmetric H_2–air cases, the situation is more confused. For the tests of Eggers (1971), the TKE and $K\varepsilon 1$ and $K\varepsilon 2$ models perform essentially equally. However, for the data of Chriss (1968), the relative performance of both $K\varepsilon$ models comes out poorer as shown in Fig. 8-32. Looking at the whole of the comparisons of predictions and experiment that are available, including wake cases, some observers see an advantage to the two-equation models for free shear flows.

8-6 ANALYSES BASED ON THE REYNOLDS STRESS

Models at this level were described in Sec. 7-12. Here we will only present some comparisons of predictions with experiment.

The predictions of the Hanjalic–Launder model are compared with data for a planar jet with $U_e = 0$ in Fig. 8-33. From that comparison and only the one other given by Launder et al. (1971), no clear advantage over the two-equation models is apparent.

The results of the Donaldson (1971) model for Reynolds and normal

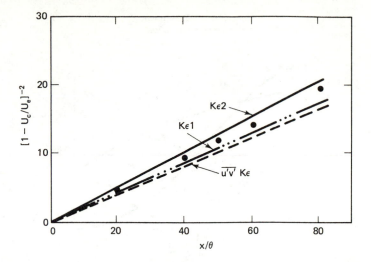

Figure 8-33 Comparison of predictions with two-equation and Reynolds stress models for the planar jet experiment of Everitt and Robins, 1978; calculations from Launder et al., 1971.

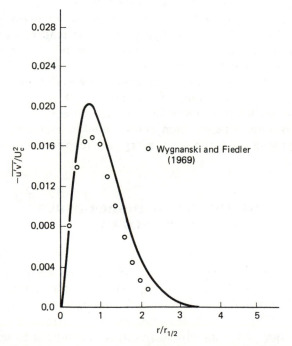

Figure 8-34 Comparison of a prediction with a Reynolds stress model and experiment for an axisymmetric jet with $U_e = 0$ for Reynolds stress. (From Donaldson, 1971.)

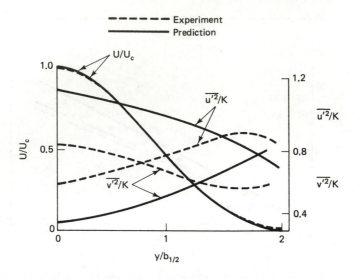

Figure 8-35 Comparison of predicion with the Daly–Harlow Reynolds stress model and experiment for a planar jet after Rodi.

stresses are compared with experimental data for a round jet with $U_e = 0$ in Fig. 8-34. Some adjustments to the many constants involved were made to obtain the results shown in these figures, and the reader is referred to the original paper for all the details. Nonetheless, good results are obviously achievable with this model.

The Daly–Harlow model has been applied by Rodi to the planar jet with $U_e = 0$, and the results are compared to experiment in Fig. 8-35. The mean flow is predicted well, but the individual normal stresses are seen to be predicted poorly. This is the same situation as was found in boundary layers (see Sec. 7-12). The shear stress and turbulence energy are reported to have been predicted accurately.

8-7 COMMENTS ON THE USE OF HIGHER-ORDER MODELS FOR FREE MIXING FLOWS

The discussion in Sec. 7-14 applies here as well. With the higher-order models, one needs boundary conditions on the additional variables as well as the usual requirements for (U, V) [and maybe $(\bar{T}, C)$]. Often, this is a serious problem. For example, in Forstall and Shapiro (1950) and Chriss (1968), no data for K or any other turbulence data are reported. It, therefore, was necessary to *estimate* the values of the initial and boundary conditions to use the higher-order models for those cases. The whole matter is discussed in some detail in Launder et al. (1971).

PROBLEMS

8.1. Derive the expression for the far-field decay of the centerline velocity for a planar jet with $U_e = 0$.

8.2. Derive the expressions for width growth and velocity defect decay in the far wake behind an axisymmetric body assuming that the profiles are *similar*.

8.3. A vertical pipe 8 m in diameter and 2000 m long is to be hung down from the surface in the Gulf Stream for an experimental power plant. If the local current is 1 m/s and constant with depth, how wide will the disturbance be at a distance of 0.5 k? How much velocity variation across the disturbance at that location can be expected? What is the value of the mixing length?

8.4. Consider the flow of air across a long, streamlined ($C_D = 0.15$) tower. The air velocity is 18 m/s and the tower has a maximum thickness of 25 cm. Could you detect the presence of a wake with a Pitot-static tube (measures total and static pressure) at a distance of 500 m? How wide would the wake be? If the strut were heated, how would the *thermal* wake compare in size to the *velocity* wake? If the strut were porous and CO_2 were slowly injected, how would the downstream pattern of concentration compare to the velocity and temperature fields? How about H_2 injection?

8.5. An experiment and an analysis are to be compared to determine a turbulent Prandtl number. Hot air (50°C) is injected at 3 m/s through a two-dimensional jet (total height = 3 cm) into a room (25°C). The analysis gives the result

$$\frac{\bar{T}(x, y) - T_e}{\bar{T}_c(x) - T_e} = \frac{1 + \cos(\pi y / 5\alpha_T x)}{2}$$

where α_T is the turbulent thermal diffusivity with units m^2/s and $T_c(x) \sim 1/x$. You

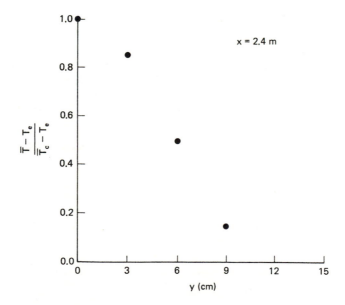

may assume that the eddy viscosity is constant throughout the flow at a value of 0.005 m²/s. Using the *data* shown in the accompanying figure, what is the Prandtl number?

8.6. In the upper layers of the ocean, wave action and thermal mixing produce a turbulent background with an eddy viscosity of approximately $\mu_T \approx 10^2$ P. If nuclear power plant waste in a seawater carrier is released through a slit into such a medium where the current is 0.5 m/s, how fast will the maximum concentration decay with downstream distance?

9

WALL-BOUNDED
TURBULENT FLOWS
WITH VARIABLE DENSITY
AND HEAT AND MASS TRANSFER

9-1 INTRODUCTION

This chapter is concerned with cases where there are differences between the wall and free-stream temperatures or injection of a foreign fluid so as to produce heat and mass transfer and/or substantial variations in fluid density and laminar, thermophysical properties. The laminar thermophysical properties remain important in wall-bounded turbulent flows because of the *laminar sublayer*. The organization of this chapter will follow that used in the earlier sections of this book on turbulent flows—experimental data for mean and fluctuating flow quantities, transport modeling at various levels, and then comparisons of predictions with experiment.

9-2 EXPERIMENTAL INFORMATION

9-2-1 Mean Flow Data

For mean flow profiles, one obviously has to deal with at least one new variable, the mean temperature $\bar{T}$. With foreign fluid injection, it is also necessary to treat the *mean* concentration C. It is reasonable to ask if there exist inner and outer region scaling laws for mean temperature profiles such as were found for the mean velocity profiles in Sec. 7-3-1. For the inner, wall-dominated regions, such an effort has been pursued for low-speed, constant-property flows by a number of workers, with the most comprehensive treatment being that by Kader (1981). The key idea is that of a *heat transfer temperature,*

$$T_* \equiv \frac{q_w}{\rho c_p u_*} \tag{9-1}$$

which corresponds to the *friction velocity* u_*. This leads to a dimensionless temperature

$$T^+ \equiv \frac{T_w - \bar{T}}{T_*} \tag{9-2}$$

that corresponds to u^+. This in turn will permit the development of a *temperature law of the wall,*

$$T^+ = g_T(y^+, \text{Pr}) \tag{9-3}$$

where Pr is the laminar Prandtl number. For the flow in the *laminar sublayer,* the same kind of development as used for the velocity field gives

$$T^+ = y^+ \text{Pr} \tag{9-4}$$

There is also a logarithmic region described by

$$T^+ = \frac{1}{\kappa_T} \ln(y^+) + C_T(\text{Pr}) \tag{9-5}$$

with

$$C_T(\text{Pr}) = (3.85\text{Pr}^{1/3} - 1.3)^2 + 2.12 \ln(\text{Pr}) \tag{9-6}$$

Kader (1981) further developed functions to join these two regions smoothly and also for the outer region corresponding to the velocity defect law. His final, complete equation is

$$T^+ = \text{Pr} \cdot y^+ \exp(-\Gamma)$$
$$+ \left[2.12 \ln \left((1 + y^+) \frac{2.5(2 - y/\delta)}{1 + 4(1 - (y/\delta))^2} \right) + C_T(\text{Pr}) \right] \exp\left(-\frac{1}{\Gamma}\right) \tag{9-7}$$

where

$$\Gamma \equiv \frac{0.01(y^+ \cdot \text{Pr})^4}{1 + 5y^+ \cdot \text{Pr}^3} \tag{9-8}$$

The success of this representation for fluids with a wide range of laminar Prandtl number is shown in Fig. 9-1. In addition, Kader (1981) gives a corresponding law for pipe and channel flows.

A useful result can be deduced with the aid of the logarithmic laws for both temperature and velocity [Eqs. (9-5) and (7-7b)] extended out to the free stream:

$$T_e^+ = \frac{1}{\kappa_T} \ln(\delta_T^+) + C_T(\text{Pr}) \tag{9-9}$$

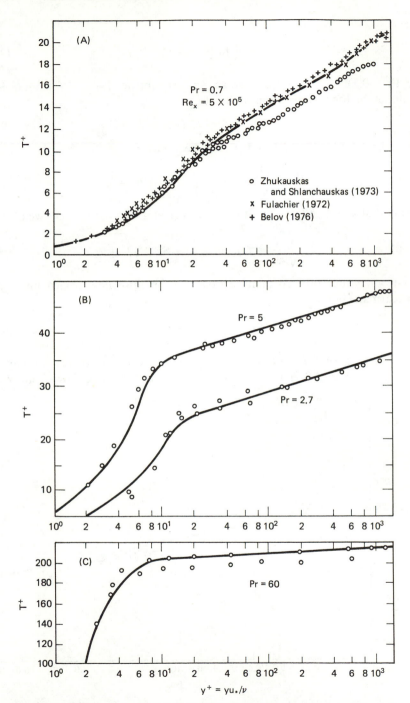

Figure 9-1   Comparison of Eq. (9-7) and data for the temperature law of the wall. (From Kader, 1981.)

and

$$u_e^+ = \frac{1}{\kappa} \ln (\delta^+) + C \tag{9-10}$$

For $\mathrm{Pr} = 0(1)$, $\delta_T \approx \delta$ and $\kappa_T \approx \kappa$, so we can subtract Eq. (9-10) from (9-9) to eliminate the logarithmic terms and get

$$T_e^+ - u_e^+ = C_T(\mathrm{Pr}) - C \tag{9-11}$$

Using the definitions of T^+ and u^+, this becomes

$$\frac{\rho c_p u_*(T_w - T_e)}{q_w} - \frac{U_e}{u_*} = C_T(\mathrm{Pr}) - C \tag{9-12}$$

Noting that $h = q_w/(T_w - T_e)$, $\mathrm{St} = h/(\rho U_e c_p)$, and $C_f = 2(u_*/U_e)^2$, we obtain

$$\sqrt{\frac{2}{C_f}} + C_T(\mathrm{Pr}) - C = \mathrm{St}^{-1} \sqrt{\frac{C_f}{2}} \tag{9-12a}$$

This result is not tidy, so many people use the equivalent approximation of von Kármán (1939):

$$\mathrm{St} = \frac{C_f}{2\mathrm{Pr}^{0.4}(T_w/T_e)^{0.4}} \tag{9-13}$$

which is good for $0.7 \leq \mathrm{Pr} \leq 10.0$. The prediction of Eq. (9-13) is compared with data in Fig. 9-2. The corresponding formula for pipe flow from White

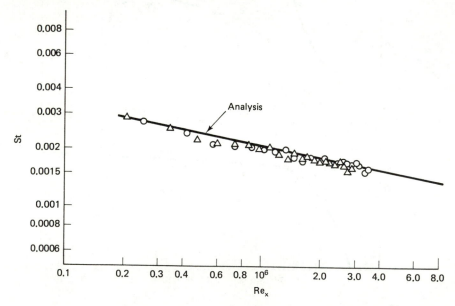

Figure 9-2 Dimensionless heat transfer coefficient for turbulent flow over a constant-temperature flat plate in low-speed flow. (From Reynolds et al., 1958.)

(1974) is

$$St = \frac{\lambda/8}{1 + 12.8(Pr^{0.68} - 1)\sqrt{\lambda/8}} \qquad (9\text{-}14a)$$

Equation (9-14a) is sometimes approximated as

$$St = 0.027(Re_D)^{-0.2}Pr^{-2/3} \qquad (9\text{-}14b)$$

This also compares well with measurements, some of which are given in Fig. 9-3 for the whole range of laminar, transitional, and turbulent flow. In Fig. 9-3, μ_b stands for the laminar viscosity evaluated at the *bulk* temperature [see Eq. (4-30)].

Most of the consideration of flows with surface mass transfer has been in terms of the analogy between heat and mass transfer introduced earlier for laminar flows. Within that framework, one presumes a direct correspondence between a film coefficient for heat h and for diffusion h_D usually presented in dimensionless terms as Nusselt numbers Nu and Nu_{Diff} or Stanton numbers St and St_{Diff}. The subscripts D and Diff denote flows with diffusion (i.e., mass transfer). Further, the role of the Prandtl number Pr in heat transfer correlations is taken by the Schmidt number Sc for mass transfer. The success of this approach can be judged by studying the data collected by Deissler (1955) for fully developed flow in pipes shown in Fig. 9-4. Note the huge range in laminar Prandtl and Schmidt numbers covered.

The agreement shown in Fig. 9-4 between mass transfer data and a transformed heat transfer law ($St \to St_{Diff}$ with $Pr \to Sc$), where the heat transfer law is for flow over a solid surface is only obtained for low surface mass transfer rates. The basic analogy between heat and mass transfer still holds for nonnegligible surface mass transfer, but in such cases one must begin with a heat transfer law for flow over a porous surface with injection or suction. The influence of injection or suction on heat transfer is substantial as was shown earlier for laminar flow and now in Fig. 9-5 for turbulent flow.

The study of turbulent flows with variable composition has not progressed to the same point as for the mean velocity and temperature fields. Thus, there are not well-developed equivalents of the *law of the wall* and *defect law* for species profiles.

For variable-density and variable-property flows produced directly by high free-stream velocities, the *law of the wall* for velocity developed for low-speed, constant-density, and constant-property flows no longer holds as shown in Fig. 9-6 from Lee et al. (1969). In this case

$$y^+ \equiv \frac{\rho_w y u_*}{\mu_w} \quad \text{and} \quad u_* \equiv \sqrt{\frac{\tau_w}{\rho_w}} \qquad (9\text{-}15)$$

since density and properties vary with y. There have been a number of attempts to develop a law suitable for high-speed cases, and those usually referred to as *Van Driest I* and *II* from Van Driest (1951) and (1956b) are

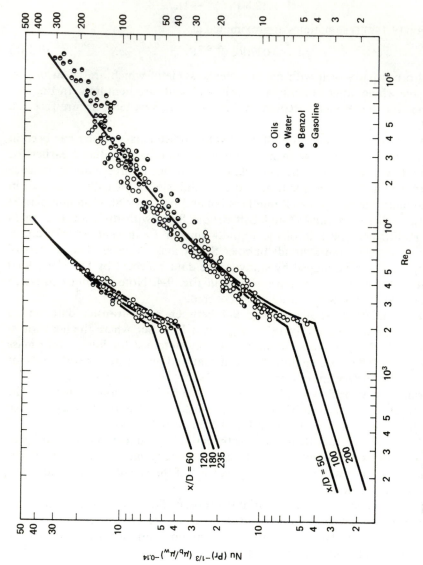

Figure 9-3 Laminar, transitional, and turbulent heat transfer in pipes. (From Sieder and Tate, 1936.)

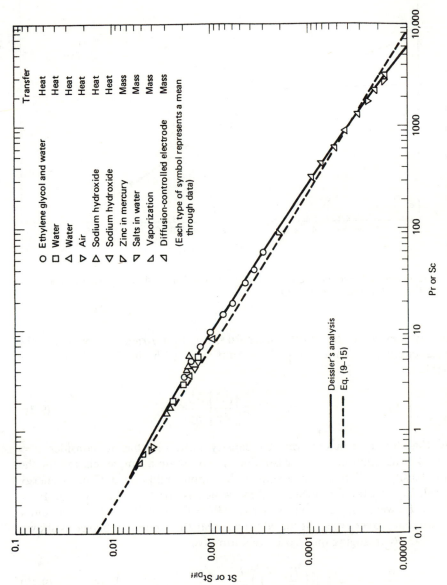

Figure 9-4 Data for heat and mass transfer in pipes ($Re_D = 10,000$) to show the close relationship between the two processes. (Collected by Deissler, 1955.)

		Transfer
○	Ethylene glycol and water	Heat
□	Water	Heat
△	Water	Heat
▽	Air	Heat
△	Sodium hydroxide	Heat
▽	Sodium hydroxide	Heat
▷	Zinc in mercury	Mass
▽	Salts in water	Mass
◁	Vaporization	Mass
◁	Diffusion-controlled electrode	Mass

(Each type of symbol represents a mean through data)

—— Deissler's analysis
- - - Eq. (9-15)

Pr or Sc

St or St$_{Diff}$

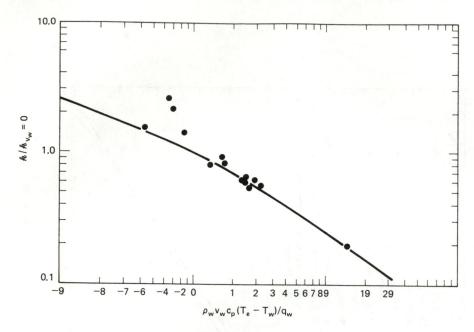

Figure 9-5 Influence of injection or suction on the film coefficient for flow over a flat plate. (From Mickley et al., 1954.)

generally viewed as the most successful. For *Van Driest I*, one takes the mixing-length formulation, Eq. (7-37), but now with the fluid density as a variable

$$\tau_T = \bar{\rho}\, \ell_m^2 \left|\frac{\partial U}{\partial y}\right| \frac{\partial U}{\partial y} \tag{9-16}$$

Note that for a turbulent, variable-density flow, one has to consider $\rho = \bar{\rho} + \rho'$. The assumption $\ell_m = \kappa y$ is retained. To obtain a simple relation for the mean density $\bar{\rho}$, Van Driest assumed the applicability of a Crocco energy integral, Eq. (5-21a), for turbulent flow, which is strictly valid only for $\text{Pr} = \text{Pr}_T = 1$. However, for air and most gases $\text{Pr} \approx 0.7$ and Pr_T is also near unity in a boundary layer except at the edge and near the laminar sublayer as shown in Fig. 9-7. Rearranging the Crocco relation gives

$$\frac{\rho_w}{\bar{\rho}} = \frac{\bar{T}}{T_w} = 1 + \left(\frac{T_{aw}}{T_w} - 1\right)\frac{U}{U_e} - \left(\frac{\gamma-1}{2}\right)M_e^2\left(\frac{U}{U_e}\right)^2\frac{T_e}{T_w} \tag{9-17}$$

Substituting Eq. (9-17) into (9-16), making the usual assumption $\tau + \tau_T = \tau_w$ in the inner region and neglecting the laminar sublayer, Van Driest solved Eq.

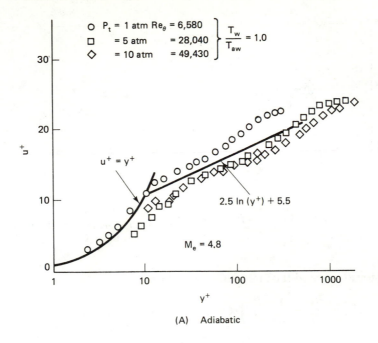

(A)   Adiabatic

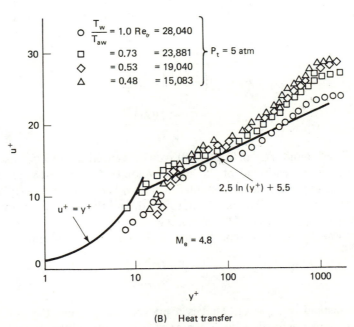

(B)   Heat transfer

Figure 9-6 Law of the wall plot of high-speed data and low-speed equations. (From Lee et al., 1969.)

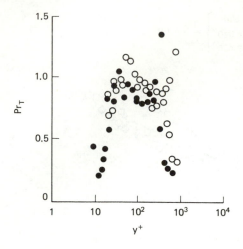

Figure 9-7 Variation of the turbulent Prandtl number in the boundary layer. (From Blom, 1970.)

(9-16) for dU/dy and then integrated to obtain

$$\frac{U_e}{B_1}\left[\sin^{-1}\left(\frac{2B_1^2(U/U_e) - B_2}{\sqrt{B_2^2 + 4(B_1)^2}}\right) + \sin^{-1}\left(\frac{B_2}{\sqrt{B_2^2 + 4(B_1)^2}}\right)\right]$$

$$= u_*\left[\frac{1}{\kappa}\ln(y^+) + C\right] \qquad (9\text{-}18)$$

with

$$B_1 = \sqrt{\left(\frac{\gamma - 1}{2}\right)M_e^2\left(\frac{T_e}{T_w}\right)}$$

$$B_2 = \frac{1 + ((\gamma - 1)/2)M_e^2}{T_w/T_e} - 1$$

The whole left-hand side of Eq. (9-18) can be viewed as an *effective velocity*, U_{eff}. If U_{eff} is used instead of the actual velocity U, Eq. (9-18) takes on the same form as the low-speed, constant-density, and constant-property *law of the wall*. A skin friction law can also be developed, and the result is compared with data as $C_f/(C_f)_{M_\infty \approx 0}$ versus M_∞ in Fig. 9-8.

For *Van Driest II*, the von Kármán (1931) mixing-length model

$$\ell_m = \kappa\left|\frac{\partial U/\partial y}{\partial^2 U/\partial y^2}\right| \qquad (9\text{-}19)$$

was used instead of $\ell_m = \kappa y$. That is a bit surprising, since the von Kármán model has never gained wide acceptance. In addition, the temperature relationship is a *modified* Crocco integral to account for cases where the recovery factor r is not unity. Using the resulting velocity profile in the momentum

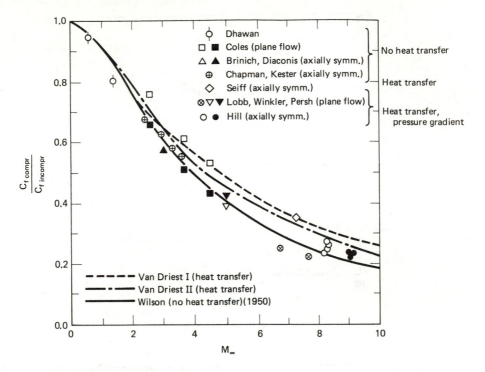

Figure 9-8   Variation of skin friction coefficient for turbulent flow over a flat plate as a function of Mach number at $\mathrm{Re}_x = 10^7$. (From Hill, 1956.)

integral equation for a flat plate,

$$\frac{d\theta}{dx} = \frac{C_f}{2} \tag{9-20}$$

a skin friction law is determined as

$$\frac{0.242}{B_1'\sqrt{C_f(T_w/T_e)}}\left[\sin^{-1}\left(\frac{2(B_1')^2 - B_2'}{\sqrt{(B_2')^2 + 4(B_1')^2}}\right) + \sin^{-1}\left(\frac{B_2'}{\sqrt{(B_2')^2 + 4(B_1')^2}}\right)\right]$$

$$= 0.41 + \log\left(\mathrm{Re}_x C_f\right) - \omega \log\left(\frac{T_w}{T_e}\right) \tag{9-21}$$

where

$$B_1' = \sqrt{\frac{((\gamma - 1)/2)M_e^2 r}{T_w/T_e}}$$

$$B_2' = \frac{1 + ((\gamma - 1)/2)M_e^2 r}{T_w/T_e} - 1$$

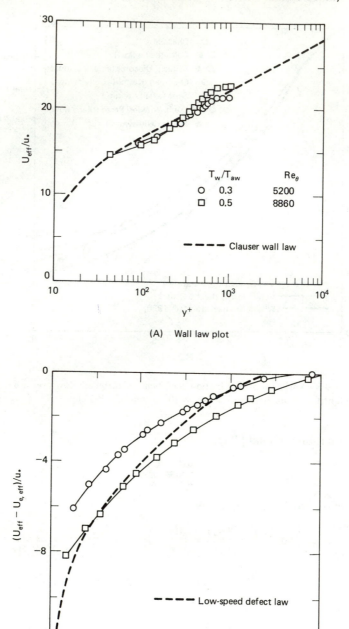

(A)   Wall law plot

(B)   Defect law plot

Figure 9-9 Comparison of correlation of velocity profiles by Van Driest II and Mach 7 data. (From Hopkins et al., 1972.)

	Wind tunnel	M_∞	Type of cone		Wind tunnel	M_∞	Type of cone
●	Ames 1 × 3 ft No. 1	1.97	10° hollow; steel	⊙	Aberdeen	2.18	10° wood
□	Lewis 8 × 6 ft	1.98	10° hollow; steel	X	GALCIT 5 × 5 in.	6.0	20° ceramic
△	Lewis 18 × 18 in.	1.94	10° hollow; steel	○	Ames 1 × 3 ft No. 1	2.0	20° hollow; steel
▽	Ames 6 × 6 ft	1.9	10° hollow; steel	▲	JPL 18 × 20 in.	4.50	5° fiberglass
◇	Ames 10 × 14 in.	4.48	10° hollow; steel	⊡	JPL 18 × 20 in.	1.63	13° Lucite
▼	Ames 2 × 2 ft	1.21	10° hollow; steel	+	JPL 18 × 20 in.	4.50	13° Lucite
⊕	Lewis 2 × 2 ft	3.93	10° hollow; steel	◬	JPL 12 × 12 in.	1.63	13° Lucite
◆	Ames 1 × 3 ft No. 2	3.00	10° hollow; steel	*	JPL 12 × 12 in.	2.54	13° Lucite

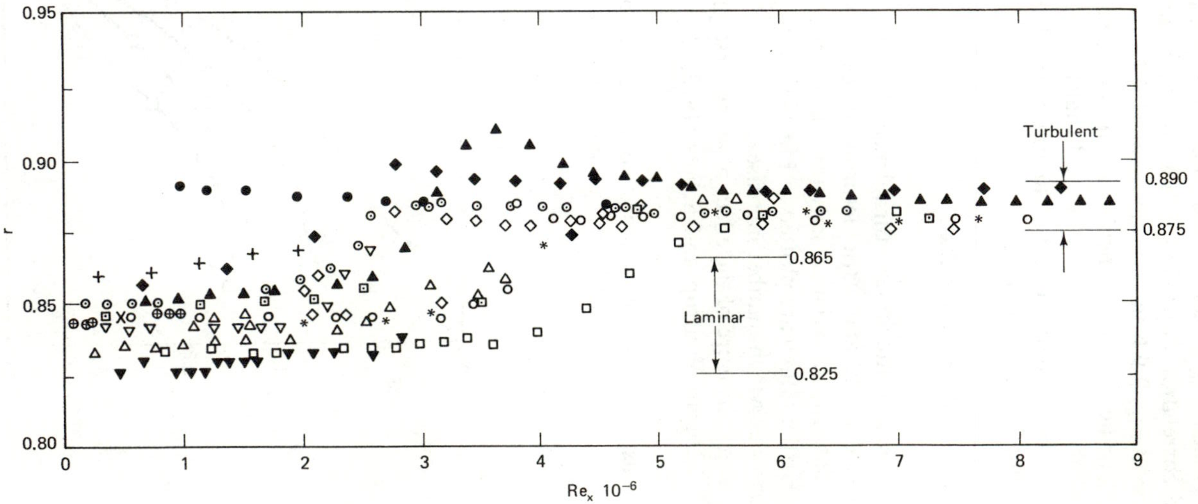

Figure 9-10 Recovery factors on cones from Mach 1.2 to 6.0. (From Mack, 1954.)

and ω comes from the assumption $\mu \sim T^{\omega}$. The improved agreement of this prediction compared with experiment can be seen in Fig. 9-8.

When U_{eff} [the left-hand side of Eq. (9-18)] is defined using B'_1 and B'_2, good correlation of profiles in both the inner and outer regions is obtained as shown in Fig. 9-9 from Hopkins et al. (1972).

For heat transfer calculations in high-speed flow, one needs to calculate the adiabatic wall temperature, and that requires information on the recovery factor r. Some data are given in Fig. 9-10. The approximation $r \approx \text{Pr}^{1/3}$ for turbulent flow is often used.

9-2-2 Turbulence Data

There is much less turbulence data in the literature for high-speed, variable-density, and variable-property cases than for the low-speed cases. Some data for the axial velocity turbulence intensity at various Mach numbers from Kistler (1959) are plotted in Fig. 9-11. One can see that the intensity level decreases as Mach number increases.

Some measure of the effect of an adverse pressure gradient can be seen in Fig. 9-12 from Waltrup and Schetz (1973). Note the change in profile shape near the wall.

Lastly, we show data for total temperature fluctuations from Kistler (1959) in Fig. 9-13. Again, the intensity decreases with increasing Mach number.

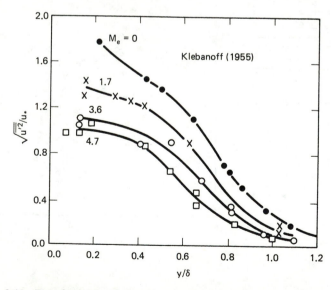

Figure 9-11 Data for axial turbulent intensity distribution in the boundary layer at various Mach numbers. (From Kistler, 1959.)

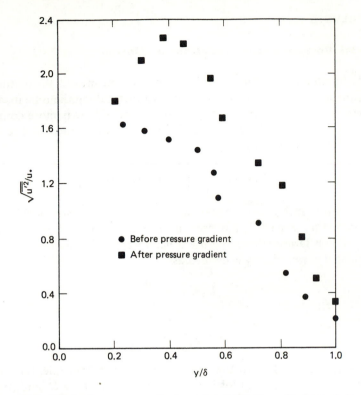

Figure 9-12 Effect of pressure gradient on the distribution of axial turbulence intensity in the boundary layer at supersonic speed. (From Waltrup and Schetz, 1973.)

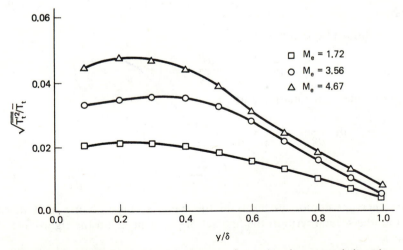

Figure 9-13 Distribution of total temperature fluctuations in supersonic boundary layers. (From Kistler, 1959.)

9-3 ANALYSIS

9-3-1 Boundary Layer Equations of Motion

The matter of splitting a variable into its mean and fluctuating parts, substituting into the equations of motion and then taking the time mean of the equations to produce equations governing the mean flow is more complicated for a variable density fluid. If one follows the procedure of Sec. 7-5 for constant-density flows using

$$U(x, y) = \frac{1}{T_0} \int_0^{T_0} u(x, y, z, t) \, dt \tag{1-19}$$

additional terms such as $\overline{\rho' u'}$ appear, since the varying density now has a mean $\bar{\rho}$ and fluctuating ρ' part. It has proven more convenient, therefore, to follow the lead of Van Driest (1951) and use a *mass-weighted averaging* procedure. A *mass-weighted mean* velocity is defined as

$$\tilde{U}(x, y) = \frac{\overline{\rho u}}{\bar{\rho}} \tag{9-22}$$

and then

$$u(x, y, z, t) = \tilde{U}(x, y) + \tilde{u}'(x, y, z, t) \tag{9-23}$$

instead of Eq. (1-20). This distinction is important in principle, but under the assumptions usually invoked for boundary layer flows, $U \approx \tilde{U}$, and so on. (See Schubauer and Tchen, 1961). The boundary layer equations of motion for steady, planar flow can then be written as

$$\frac{\partial}{\partial x}(\bar{\rho} U) + \frac{\partial}{\partial y}(\overline{\rho v}) = 0 \tag{9-24}$$

for the continuity equation and

$$\bar{\rho} U \frac{\partial U}{\partial x} + \overline{\rho v} \frac{\partial U}{\partial y} = -\frac{dP}{dx} + \frac{\partial}{\partial y}\left(\mu \frac{\partial U}{\partial y} - \overline{\bar{\rho} u' v'}\right) \tag{9-25}$$

for the x momentum equation and

$$\bar{\rho} U \frac{\partial \bar{h}}{\partial x} + \overline{\rho v} \frac{\partial \bar{h}}{\partial y} = \frac{\partial}{\partial y}\left(k \frac{\partial \bar{T}}{\partial y} - \overline{\bar{\rho} v' h'}\right)$$
$$+ \left(\mu \frac{\partial U}{\partial y} - \overline{\bar{\rho} u' v'}\right)\frac{\partial U}{\partial y} + U \frac{dP}{dx} \tag{9-26}$$

for the energy equation in terms of the mean static enthalpy $\bar{h}$. If there is diffusion, the second term in the first set of parentheses on the right-hand side becomes

$$-\sum_i \bar{\rho} \, C_i \overline{v' h_i'} \tag{9-26a}$$

and two additional terms for heat transport via mass transfer must be added:

$$\frac{\partial}{\partial y}\left(\sum_i \bar{h}_i \left(\bar{\rho} \, D_{12} \, \frac{\partial C_i}{\partial y} - \overline{\rho v' c_i'}\right)\right) \tag{9-26b}$$

Also, a species conservation equation is then required:

$$\bar{\rho}U \, \frac{\partial C_i}{\partial x} + \overline{\rho v} \, \frac{\partial C_i}{\partial y} = \frac{\partial}{\partial y}\left(\bar{\rho}D_{12} \, \frac{\partial C_i}{\partial y} - \overline{\rho v' c_i'}\right) \tag{9-27}$$

In Eqs. (9-25) to (9-27) it is implied that $\mu = \bar{\mu} = \mu(\bar{T})$, $k = \bar{k} = k(\bar{T})$, and $D_{12} = \bar{D}_{12} = D_{12}(\bar{T})$.

9-3-2 Mean Flow Models for the Eddy Viscosity and Mixing Length

Looking at Eqs. (9-25) to (9-27), it is clear that it is again necessary to *model* the turbulent shear, $-\overline{\rho u' v'}$, heat transfer, $\overline{\rho v' h'}$, and mass transfer, $\overline{\rho v' c_i'}$. With an *eddy viscosity formulation*, one writes

$$-\overline{\rho u' v'} = \mu_T \, \frac{\partial U}{\partial y} \tag{9-28}$$

and with a *mixing-length formulation*

$$-\overline{\rho u' v'} = \bar{\rho} \, \ell_m^2 \left|\frac{\partial U}{\partial y}\right| \frac{\partial U}{\partial y} \tag{9-29}$$

Turbulent heat transfer is almost universally treated simply via a constant *turbulent* Prandtl number $\text{Pr}_T = \mu_T c_p / k_T$ [see Eq. (8-4)]. Similarly, turbulent mass transfer calculations are usually based on a constant *turbulent* Schmidt number $\text{Sc}_T = \mu_T/(\rho D_T)$ [see Eq. (8-7)].

One might have expected that the derivation of mean flow models for the inner and outer regions for variable-density flows would follow the procedure used for constant-density flows. Recall, in that case, that the procedure started with the *law of the wall* and the *defect law* as *given*, and models were found that, when used in the equations of motion, would reproduce those laws (see Sec. 7-8). An equivalent development for variable density flows would start with a *law of the wall* and a *defect law* for variable-density boundary layers and then proceed along the same lines as before. Perhaps surprisingly, that was *not* the procedure followed. Indeed, the current models in use come about as ad hoc extensions of the constant-density models.

For the inner region, Patankar and Spalding (1967) and Cebeci (1971) simply used the Van Driest model [Eq. (7-63)] with the density and viscosity evaluated at the local conditions as a function of y in the inner region. Anderson and Lewis (1971) did the same thing with the Reichardt model [Eq. (7-64)]. Corresponding mixing-length models can be found with Eq. (7-40). No soundly based analysis has been presented to justify these assumptions. The justification is the pragmatic argument that the models seem to *work*. By that one means that predictions obtained by using the subject model in a numerical solution procedure agree within an acceptable level with data for a few specific cases.

For the outer region, Maise and McDonald (1968) have analyzed the available data to conclude that the mixing-length model, Eq. (7-73), can be carried over directly to variable-density cases. That assumption was used in the computation method of Patankar and Spalding (1967).

Those workers that favor the Clauser model for the eddy viscosity in the outer region, Eq. (7-85), have also tried to extend it to variable-density cases on a purely ad hoc basis. Herring and Mellor (1968) and Cebeci (1971) simply assert that the Clauser model should be extended to variable-density flows as

$$\mu_T = C\bar{\rho}U_e\,\delta_k^*$$

$$= C\bar{\rho}U_e \int_0^\delta \left(1 - \frac{U}{U_e}\right) dy \tag{9-30}$$

rather than using the *true* displacement thickness for variable-density flows

$$\delta^* \equiv \int_0^\delta \left(1 - \frac{\bar{\rho}U}{\rho_e U_e}\right) dy \tag{9-31}$$

It is important to emphasize that Eq. (9-30) was not rederived for variable-density flows using an appropriate *defect law* following the development of Clauser. The justification is again that the users feel it seems to *work*. The form adopted in Eq. (9-30) is, however, contrary to that found appropriate for free mixing flows with large density variations (see Sec. 8-3-2). For free mixing flows, the Clauser model extended on the basis of the true δ^*, not δ_k^*, was successful. The author has tested the use of δ_k^* for free mixing flows and shown it to produce poor predictions compared to those with the true δ^*. According to the two-layer view of turbulent boundary layers, the outer region of a boundary layer can be closely likened to a free mixing shear flow. This writer feels that this apparent contradiction can be partially resolved by noting that most turbulent boundary layers that have been considered for comparisons between prediction and experiment have much smaller density variations than the free mixing flows that have been considered (e.g., the H_2–air jet mixing cases discussed in Sec. 8-3-2, where $\bar{\rho}_j/\bar{\rho}_e \approx \frac{1}{15}$). Thus, the distinction between δ_k^* and δ^* is much less important in most boundary layer flows. To help clarify the matter, some comparisons with both representations will be shown below. This question cannot, however, be considered closed, and further detailed study is warranted.

We are now in a position to discuss comparisons between experiment and predictions based on the various mean flow models for variable-density boundary layer flows. One such comparison was presented in Fig. 5-15 from Anderson and Lewis (1971), who used the extended Reichardt model and Eq.

(9-30). Cebeci and coworkers (see Cebeci, 1971) have produced a number of comparisons using the extended Van Driest model and Eq. (9-30), with generally good results. Herring and Mellor (1968) use their own inner region model for both constant- and variable-density cases, which is essentially a mixing-length model and Eq. (9-30), and again obtain generally good results.

Schetz and Favin (1971) used the extended Reichardt model for the inner region and compared predictions based on both δ^* and δ_k^* with experiment. Some results are given in Figs. 9-14 and 9-15 for supersonic flows without and with wall heat transfer. Clearly, based on these results it is hard to argue convincingly for δ_k^* over δ^* (or for that matter, vice versa).

Schetz and Favin (1971) also treated boundary layer flows with light gas injection through a porous wall such that sizable density variations are produced even at low-speed conditions. The procedure used the Reichardt model extended to injection, Eq. (7-71), with local $\bar{\rho}$ and μ for the inner region and the Clauser model with both δ^* and δ_k^* (for comparison) for the outer region. Calculations were performed to compare the predictions of the analysis with the data of Scott et al. (1964) for helium injection. Figure 9-16(A) shows some results in terms of the skin friction coefficient and the integral boundary layer thicknesses. There is not a great deal of difference between the results for the two methods; both give predictions that are low for the momentum thickness and C_f, with somewhat better agreement for the δ^* model, and both agree well with the displacement thickness data. The velocity profile measured at the farthest downstream station is compared to the theory in Fig. 9-16(B); the agreement is quite good for both models but is better for the δ^* model near the wall and at the outer edge of the layer. The experimental measurements were interpreted by Scott et al. (1964) to provide the turbulent eddy viscosity distribution across the layer. A comparison with the eddy viscosity distributions obtained numerically using δ^* and δ_k^* in the outer region model is shown in Fig. 9-16(C). In the outer region, the δ^* model is better. However, at large y/δ, the theoretical models deviate considerably, suggesting that the Klebanoff *intermittency factor* used falls off too rapidly toward the outer edge of the layer.

Based on all the results in the literature for boundary layers and those in Chap. 8 for free mixing flows, the author suggests the use of the Clauser model extended directly using δ^*, not δ_k^*, or the mixing-length model, $\ell_m \approx 0.09\delta$, for the outer region. For the inner region, the Reichardt model with local $\bar{\rho}$ and μ is most convenient in numerical treatments.

9-3-3 Methods Based on Models for the Turbulent Kinetic Energy

Variable-density versions of both approaches to TKE models (the Prandtl energy method and the Bradshaw model) have been developed. The modeling of the various terms in the TKE equation is as already described in Sec. 7-10.

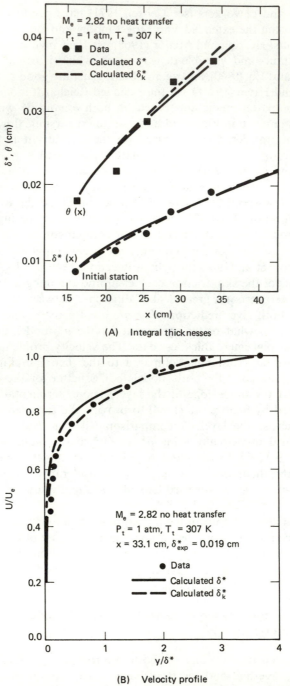

Figure 9-14 Comparison of predictions with mean flow models with the data of Monaghan and Cooke (1952) for adiabatic flow. (From Schetz and Favin, 1971.)

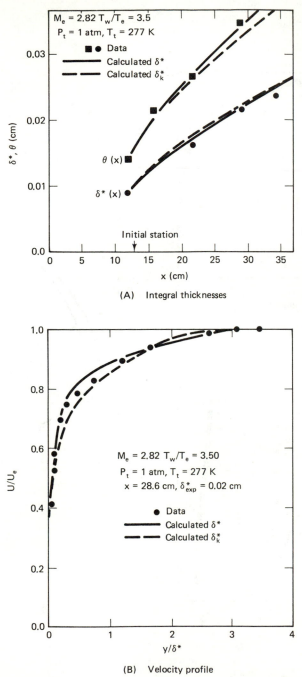

(A)   Integral thicknesses

(B)   Velocity profile

Figure 9-15   Comparison of predictions with mean flow models with the data of Monaghan and Cooke (1952) for heat transfer cases. (From Schetz and Favin, 1971.)

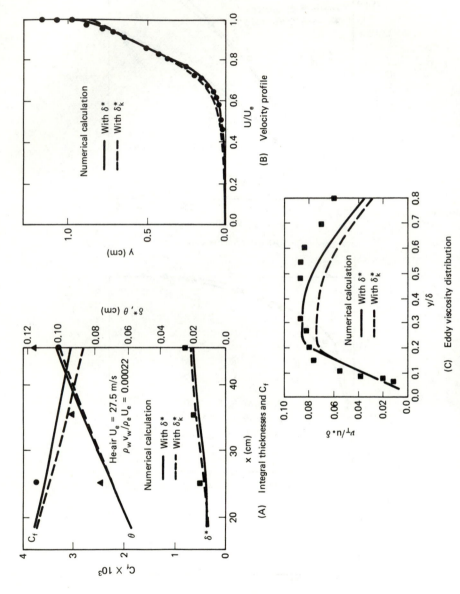

Figure 9-16 Comparison of predictions with mean flow models and the data of Scott et al. (1964) for helium injection into the boundary layer. (From Schetz and Favin, 1971.)

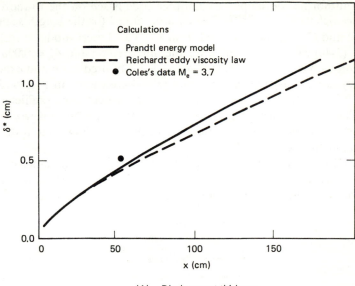

(A)   Displacement thickness

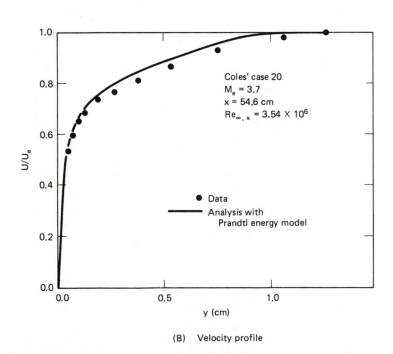

(B)   Velocity profile

Figure 9-17 Comparison of predictions with the Prandtl energy method and the supersonic flat-plate data of Coles (1954). (From Schetz et al., 1982.)

The author has developed a computer code based on the Prandtl energy method (see Schetz et al., 1982). The required model for the length scale ℓ [see Eqs. (7-90) and (7-91)] was obtained from the Van Driest model translated to ℓ_m via Eq. (7-40) with local $\bar{\rho}$ and μ for the inner region and $\ell_m = 0.09\delta$ for the outer region with $\ell = C_D^{1/4}\ell_m$. Some comparisons of prediction and experiment for the supersonic flat-plate case of Coles (1954) are shown in Fig. 9-17. Coles reported $C_f = 0.00162$ at this station, and the calculation predicted $C_f = 0.00177$. These results clearly agree well with experiment, but it is worth noting that the eddy viscosity methods discussed in the preceding section do as well (see Anderson and Lewis, 1971).

Bradshaw has developed a code incorporating a variable-density extension to his method. Sivasegaram and Whitelaw (1971) made numerous comparisons between the predictions and data for adiabatic wall cases. They also

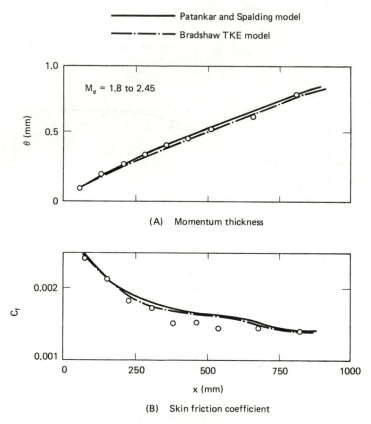

Figure 9-18 Comparison of predictions with the mixing-length model of Patankar and Spalding and the Bradshaw TKE model with supersonic favorable pressure gradient data. (From Sivasegaram and Whitelaw, 1971.)

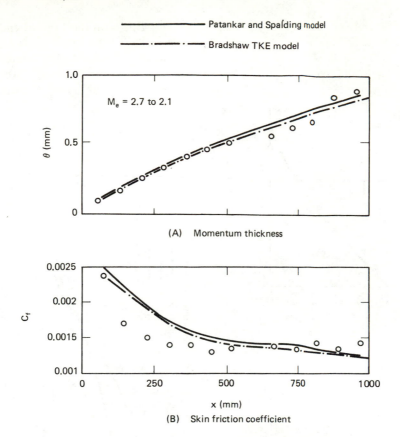

Figure 9-19 Comparison of predictions with the mixing-length model of Patankar and Spalding and the Bradshaw TKE model with supersonic adverse pressure gradient data. (From Sivasegaram and Whitelaw, 1971.)

used the mixing-length method of Patankar and Spalding (1967). Two comparisons are shown in Figs. 9-18 and 9-19. A few points are clear from these results. First, there is not a significant difference between the results of the mixing-length or TKE models. Second, predictions of adequate accuracy for design are obtainable. Finally, the poorest accuracy is found for adverse pressure gradient cases.

9-3-4 Methods Based on Models for Turbulent Kinetic Energy and a Length Scale

The $K\varepsilon$ models, for example, have not gained wide popularity for variable-density, wall-bounded flows. The method of Jones and Launder (1972) can be suggested as a promising candidate for such an extension.

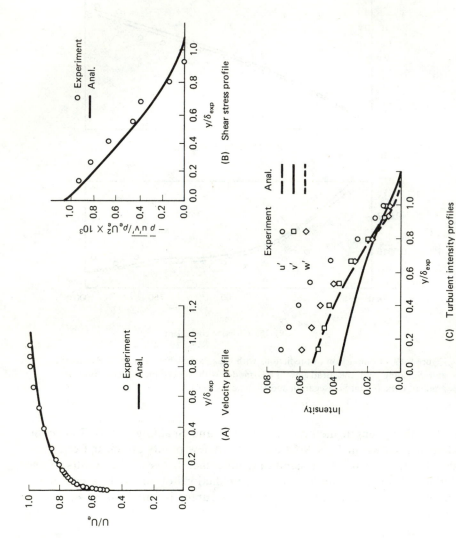

Figure 9-20 Comparison of predictions with the Donaldson Reynolds stress model and the Mach 0.6 data of Acharya (1976). (From Rubesin et al., 1977.)

9-3-5 Methods Based Directly on the Reynolds Stress

The model of Donaldson (1971) (see Secs. 7-12 and 8-6) has been applied by Rubesin et al. (1977) to various cases, including the subsonic, compressible ($M_\infty = 0.6$) experiments of Acharya (1976). Some comparisons are shown in Fig. 9-20. As for the constant-density cases (see Fig. 7-60), the mean flow, the turbulent shear, $-\bar{\rho}\overline{u'v'}$, and $\overline{v'^2}$ are predicted most accurately. However, $\overline{v'^2}$ is not predicted as well here as for constant-density cases.

On the basis of the comparisons shown in this book and others, Rubesin et al. (1977) concluded that the Reynolds stress models perform comparable to and sometimes better than the mean flow models in predicting skin friction for attached boundary layers.

9-3-6 Discussion

The observations about initial and boundary conditions at the end of Chapters 7 and 8 apply here as well.

PROBLEMS

9.1. Compare the thickness of the thermal laminar sublayer for air, water, and H_2 flow at 300 K, all at the same Reynolds number.

9.2. Consider water at 25°C flow at 10 m/s over a flat plate. What are C_f and St at $L = 1.0$ m? For a wall temperature of 20°C, what are u_* and T_*?

9.3. Consider CO_2 flowing in a tube with H_2 injection at a low rate through the walls. The pressure is 3.0 atm and the temperature is 500 K. The Reynolds number is $Re_D = 10^5$. Calculate the film coefficient for diffusion.

9.4. For flow over a 5° half-angle wedge of length 1.0 m, compare the wall shear τ_w at $M_\infty = 2.0$ and 4.0 at sea level.

9.5. Air at $M_e = 2.0$ flows over an insulated flat plate. Assuming that $U/U_e = (y/\delta)^{1/7}$ and the Crocco integral, compare δ^*/δ and δ_k^*/δ.

APPENDIX A

LAMINAR THERMOPHYSICAL PROPERTIES FOR SELECTED FLUIDS

This appendix contains laminar thermophysical property information for a few selected fluids. It is designed only to be a convenient source for the student to aid in working the homework problems. The information included was obtained from two main sources. For gases, the computer code developed by Svehla and McBride (1973) at NASA was used. For liquid water, the data compiled by the Thermophysical Properties Research Center (Touloukian and Ho, 1970, 1975) was used. Both are excellent, convenient sources of information for a wide range of fluids for the working professional.

TABLE A-1   Conversion Factors for Viscosity

Multiply by ↓ to Obtain ⟶	$N \cdot s/m^2$	$Pa \cdot s$	Poise	$lb_f \cdot s/ft^2$
$N \cdot s/m^2$	1	1	10	2.0885×10^{-2}
$Pa \cdot s$	1	1	10	2.0885×10^{-2}
Poise	10^{-1}	10^{-1}	1	2.0885×10^{-3}
$lb_f \cdot s/ft^2$	4.7880×10^1	4.7880×10^1	4.7880×10^2	1

TABLE A-2   Conversion Factors for Thermal Conductivity

Multiply by ↓ to Obtain ⟶	Btu/h ft^{-1}/R	cal/s cm^{-1}/K	J/s cm^{-1}/K	W/cm K^{-1}
Btu/h ft^{-1}/R	1	4.1338×10^{-3}	1.7296×10^{-2}	1.7296×10^{-2}
cal/s cm^{-1}/K	2.4191×10^{2}	1	4.184	4.184
J/s cm^{-1}/K	5.7818×10^{1}	2.3901×10^{-1}	1	1
W/cm K^{-1}	5.7818×10^{1}	2.3901×10^{-1}	1	1

TABLE A-3   Air at 1.0 atm

Temperature (K)	Viscosity Poise	Conductivity [cal/(cm)(s)(K)]	Specific Heat, c_p [cal/(g)(K)]	Prandtl Number
300	$184. \times 10^{-6}$	$63. \times 10^{-6}$	0.2402	0.7056
400	227.	78.	0.2422	0.7056
500	265.	92.	0.2461	0.7057
600	299.	106.	0.2512	0.7061
700	331.	121.	0.2569	0.7064
800	362.	135.	0.2626	0.7064
900	391.	148.	0.2679	0.7060
1000	419.	162.	0.2727	0.7054

TABLE A-4   Water Vapor at 1.0 atm

Temperature (K)	Viscosity Poise	Conductivity [cal/(cm)(s)(K)]	Specific Heat, c_p [cal/(g)(K)]	Prandtl Number
400	$132. \times 10^{-6}$	$72. \times 10^{-6}$	0.4547	0.8362
500	173.	100.	0.4670	0.8099
600	213.	130.	0.4814	0.7902
700	254.	163.	0.4970	0.7752
800	295.	198.	0.5133	0.7637
900	336.	230.	0.5300	0.7553
1000	376.	275.	0.5468	0.7488

TABLE A-5   Hydrogen at 1.0 atm

Temperature (K)	Viscosity Poise	Conductivity [cal/(cm)(s)(K)]	Specific Heat, c_p [cal/(g)(K)]	Prandtl Number
300	$89. \times 10^{-6}$	$439. \times 10^{-6}$	3.4208	0.6945
400	108.	539.	3.4578	0.6939
500	126.	629.	3.4709	0.6929
600	142.	713.	3.4774	0.6922
700	157.	794.	3.4891	0.6917
800	172.	874.	3.5115	0.6916
900	186.	954.	3.5442	0.6913
1000	200.	1035.	3.5810	0.6908

TABLE A-6   Carbon Dioxide at 1.0 atm

Temperature (K)	Viscosity Poise	Conductivity [cal/(cm)(s)(K)]	Specific Heat, c_p [cal/(g)(K)]	Prandtl Number
300	$150. \times 10^{-6}$	$40. \times 10^{-6}$	0.2080	0.7290
400	194.	60.	0.2242	0.7321
500	234.	77.	0.2423	0.7344
600	271.	95.	0.2572	0.7358
700	305.	112.	0.2693	0.7365
800	335.	127.	0.2793	0.7366
900	363.	142.	0.2877	0.7366
1000	389.	156.	0.2949	0.7364

TABLE A-7   Liquid Water

Temperature (K)	Viscosity Poise	Conductivity [cal/(cm)(s)(K)]	Specific Heat, c_p [cal/(g)(K)]	Prandtl Number
273.15	1.753×10^{-2}	1.349×10^{-3}	1.003	13.034
300	0.823	1.454	0.998	5.649
350	0.360	1.595	1.002	2.262
400	0.215	1.641	1.013	1.327
450	0.151	1.608	1.055	0.991
500	0.117	1.518	1.110	0.856
550	0.097	1.365	1.259	0.895
600	0.079	1.150	1.598	1.098
647[a]	0.042	—	—	—

[a] Critical temperature.

TABLE A-8   Seawater

Temperature (K)	Viscosity Poise	Kinematic Viscosity (m^2/s)
280	1.48×10^{-2}	1.490×10^{-6}
285	1.29	1.297
290	1.13	1.143
295	1.02	1.017
300	0.92	0.913

TABLE A-9   Binary Diffusion Coefficients D_{ij} (cm^2/s)

Pair	Temperature (K)							
	300	400	500	600	700	800	900	1000
$CO_2–CO_2$	0.110	0.190	0.287	0.399	0.524	0.658	0.803	0.958
$CO_2–H_2$	0.553	0.932	1.384	1.899	2.477	3.107	4.094	4.892
$CO_2–H_2O$	0.108	0.211	0.347	0.517	0.717	0.942	1.194	1.469
$CO_2–N_2$	0.174	0.293	0.436	0.600	0.782	0.982	1.034	1.233
$CO_2–O_2$	0.161	0.277	0.418	0.580	0.762	0.965	1.050	1.256
$H_2–H_2$	1.466	2.386	3.482	4.760	6.150	7.696	9.390	11.237
$H_2–H_2O$	0.886	1.502	2.237	3.080	4.024	5.057	6.179	7.386
$H_2–N_2$	0.622	1.013	1.473	1.994	2.576	3.217	3.912	4.660
$H_2–O_2$	0.914	1.490	2.170	2.944	3.800	4.745	5.772	6.876
$H_2O–H_2O$	0.088	0.202	0.373	0.603	0.891	1.235	1.636	2.086
$H_2O–N_2$	0.140	0.268	0.437	0.645	0.890	1.169	1.482	1.825
$H_2O–O_2$	0.258	0.442	0.665	0.922	1.210	1.527	1.871	2.241
$N_2–N_2$	0.204	0.335	0.490	0.665	0.861	1.075	1.309	1.561
$N_2–O_2$	0.206	0.340	0.498	0.679	0.884	1.113	1.368	1.643
$O_2–O_2$	0.206	0.341	0.502	0.686	0.894	1.126	1.381	1.656

APPENDIX B

DERIVATION
OF THE
NAVIER-STOKES EQUATIONS

There are flow situations where viscous effects are important, but the boundary layer assumptions (essentially $\partial p/\partial y = 0$) are not valid. Some of these cases were alluded to earlier when discussing separated flows. In such flows, the full viscous equations, usually called the Navier–Stokes equations, must be employed. There are both laminar and turbulent versions. Although the solution of those equations for general flows is beyond the scope of this book, it was deemed worthwhile to include a simple derivation for laminar, constant-property cases for reference.

The derivation follows along the same lines as that for the differential boundary layer equations in Chap. 3 except that now $\partial p/\partial y \neq 0$ and a more complete and more complex representation for the shear beyond $\tau = \mu(\partial u/\partial y)$ will be necessary. Consider the surface forces acting on the fluid element shown in Fig. B-1. The total surface forces acting on the faces perpendicular to the x axis are denoted as

$$\mathbf{T}_x \, dy \, dz \qquad \text{and} \qquad \left(\mathbf{T}_x + \frac{\partial \mathbf{T}_x}{\partial x} \, dx \right) dy \, dz$$

The net force from these is

$$\frac{\partial \mathbf{T}_x}{\partial x} \, dx \, dy \, dz \qquad\qquad\qquad (\text{B-1})$$

By analogy, the net forces on the other faces can be written

$$\frac{\partial \mathbf{T}_y}{\partial y} \, dy \, dx \, dz \qquad \text{and} \qquad \frac{\partial \mathbf{T}_z}{\partial z} \, dz \, dx \, dy \qquad (\text{B-2})$$

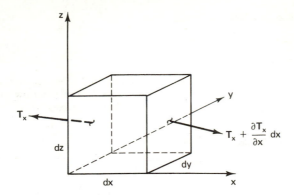

Figure B-1   General surface forces acting on the sides of a fluid element perpendicular to the x axis.

Now each of the force (per unit area, i.e., stress) vectors $\mathbf{T}_x$, $\mathbf{T}_y$, and $\mathbf{T}_z$ can be decomposed as

$$\mathbf{T}_x = \bar{i}\tau_{xx} + \bar{j}\tau_{xy} + \bar{k}\tau_{xz}$$
$$\mathbf{T}_y = \bar{i}\tau_{yx} + \bar{j}\tau_{yy} + \bar{k}\tau_{yz} \qquad \text{(B-3)}$$
$$\mathbf{T}_z = \bar{i}\tau_{zx} + \bar{j}\tau_{zy} + \bar{k}\tau_{zz}$$

The subscript notation on τ_{xy}, for example, indicates quantities acting on the face perpendicular to the x axis and in the y direction. Taking moments of the forces about an arbitrary axis and setting the result equal to zero for a fluid particle in equilibrium leads directly to the result that

$$\tau_{xy} = \tau_{yx}$$
$$\tau_{xz} = \tau_{zx} \qquad \text{(B-4)}$$
$$\tau_{yz} = \tau_{zy}$$

must hold in general.

A quantity with two subscripts (i.e., nine components) is called a *tensor*. This follows in a logical sequence from a *scalar* such as temperature, with only one component, to a *vector* such as velocity, with three components, to a stress *tensor*, with nine components. With Eq. (B-4), we see that the stress consists of a *symmetrical tensor*. The diagonal components τ_{xx}, τ_{yy}, τ_{zz} are called *normal stresses*, and they clearly bear a close relation to the pressure. The off-diagonal components are viscous *shear stresses*.

The task is now to relate the viscous stresses to the dynamics of the flow. The major physical arguments necessary have been presented in Sec. 1-3. There are also some mathematical/physical properties of the stress tensor that will prove useful. One can begin with the simple situation shown in Fig. B-2 (this is the same case as for the viscometer in Fig. 1-2) and attempt to gener-

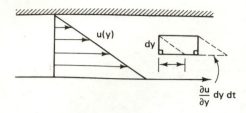

Figure B-2 Velocity distribution of a viscous fluid in a narrow gap between a moving and a fixed surface.

alize. For this simple case, we have

$$\tau = \mu \frac{\partial u}{\partial y} \tag{B-5}$$

The quantity $\partial u/\partial y$ is twice the *shear strain rate*, which is defined as one-half the rate of change of the angle between two lines in the fluid that are initially perpendicular. Thus, Eq. (B-5) can be read to say that the shear stress is proportional to the shear strain rate.

In the general case, the shear strain rate is a tensor whose components can be developed by considering Fig. B-3 with ε_{xy} as an example. Clearly,

$$\varepsilon_{xy} = \frac{1}{2}\left(\frac{d\phi_1}{dt} + \frac{d\phi_2}{dt}\right) = \frac{1}{2}\left(\frac{\partial v}{\partial x} + \frac{\partial u}{\partial y}\right) \tag{B-6}$$

and by analogy

$$\varepsilon_{yz} = \frac{1}{2}\left(\frac{\partial w}{\partial y} + \frac{\partial v}{\partial z}\right) \quad \text{and} \quad \varepsilon_{zx} = \frac{1}{2}\left(\frac{\partial u}{\partial z} + \frac{\partial w}{\partial x}\right) \tag{B-6a}$$

As in solid mechanics, this tensor is symmetric. Thus, Eq. (B-5) is generalized

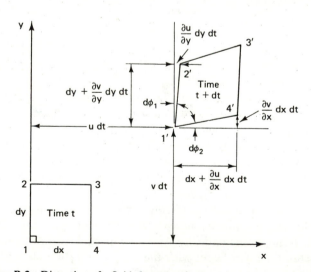

Figure B-3 Distortion of a fluid element as it moves in a viscous shear flow.

to

$$\tau_{xy} = \mu\left(\frac{\partial u}{\partial y} + \frac{\partial v}{\partial x}\right)$$

$$\tau_{yz} = \mu\left(\frac{\partial v}{\partial z} + \frac{\partial w}{\partial y}\right) \tag{B-7}$$

$$\tau_{zx} = \mu\left(\frac{\partial w}{\partial x} + \frac{\partial u}{\partial z}\right)$$

for the shear stresses. The key assumption is that the viscous stress tensor is a linear function of the rate of strain tensor. It has also been assumed that the fluid is *isotropic*; that is, the process is independent of direction.

The diagonal terms, which are the normal stresses, require special consideration. First, the pressure, p, in a fluid is usually defined as the negative (i.e., acting inward) of the mean value of the normal stresses on a sphere of unit radius, so the pressure must appear in the diagonal terms of the tensor. Also, we wish the final relation to be invariant under a simple rotation of the coordinate system. This requires the introduction of additional items in the diagonal terms of the tensor and a second viscosity coefficient, λ. The formulation is similar to solid mechanics, where we require two coefficients of elasticity for an isotropic solid. The final forms of the diagonal terms become

$$\tau_{xx} = 2\mu\frac{\partial u}{\partial x} + \lambda\left(\frac{\partial u}{\partial x} + \frac{\partial v}{\partial y} + \frac{\partial w}{\partial z}\right) - p$$

$$\tau_{yy} = 2\mu\frac{\partial v}{\partial y} + \lambda\left(\frac{\partial u}{\partial x} + \frac{\partial v}{\partial y} + \frac{\partial w}{\partial z}\right) - p \tag{B-8}$$

$$\tau_{zz} = 2\mu\frac{\partial w}{\partial z} + \lambda\left(\frac{\partial u}{\partial x} + \frac{\partial v}{\partial y} + \frac{\partial w}{\partial z}\right) - p$$

But the average of the normal stresses must be equal to the negative of the pressure whether there is motion (and therefore viscous normal stresses) or not. Thus,

$$-p = \frac{\tau_{xx} + \tau_{yy} + \tau_{zz}}{3} = -p + \frac{3\lambda + 2\mu}{3}\left(\frac{\partial u}{\partial x} + \frac{\partial v}{\partial y} + \frac{\partial w}{\partial z}\right) \tag{B-9}$$

Obviously, the additional term on the right-hand side must be zero. That is no problem for incompressible fluids, since the quantity in parentheses is zero by mass conservation [see Eq. (3-6a) for the two-dimensional form]. For other fluids, it is common to simply assume that $\lambda = -2\mu/3$. This completes the modeling of the stresses.

We turn to the fluid element in Fig. 3-1 to apply conservation of momentum and note that the pressure is now allowed to be a function of y (and z). Clearly, it will be necessary here to balance forces with momentum changes in all three coordinate directions. It is convenient to use the x direction as an example and then write down the corresponding results in the y and z directions by extension. The development of the momentum change follows directly from that in Sec. 3-3, since no approximations in those terms were made. Generalizing the development of the net surface forces acting in the x directions from that in Sec. 3-3, we obtain after dividing by the volume $(dx)(dy)(dz)$

$$\frac{\partial \tau_{xx}}{\partial x} + \frac{\partial \tau_{xy}}{\partial y} + \frac{\partial \tau_{xz}}{\partial z} \tag{B-10}$$

Using Eqs. (B-7) and (B-8) with $\lambda = -2\mu/3$, this becomes

$$-\frac{\partial p}{\partial x} + \mu\left(\frac{\partial^2 u}{\partial x^2} + \frac{\partial^2 u}{\partial y^2} + \frac{\partial^2 u}{\partial z^2}\right) + \tfrac{1}{3}\mu\frac{\partial}{\partial x}\left(\frac{\partial u}{\partial x} + \frac{\partial v}{\partial y} + \frac{\partial w}{\partial z}\right) \tag{B-11}$$

Setting the net forces equal to the change in momentum, we obtain

$$\frac{\partial u}{\partial t} + u\frac{\partial u}{\partial x} + v\frac{\partial u}{\partial y} + w\frac{\partial u}{\partial z} = f_x - \frac{1}{\rho}\frac{\partial p}{\partial x}$$
$$+ \nu\left(\frac{\partial^2 u}{\partial x^2} + \frac{\partial^2 u}{\partial y^2} + \frac{\partial^2 u}{\partial z^2}\right) \tag{B-12}$$
$$+ \tfrac{1}{3}\nu\frac{\partial}{\partial x}\left(\frac{\partial u}{\partial x} + \frac{\partial v}{\partial y} + \frac{\partial w}{\partial z}\right)$$

The corresponding equations in the y and z directions are

$$\frac{\partial v}{\partial t} + u\frac{\partial v}{\partial x} + v\frac{\partial v}{\partial y} + w\frac{\partial v}{\partial z} = f_y - \frac{1}{\rho}\frac{\partial p}{\partial y}$$
$$+ \nu\left(\frac{\partial^2 v}{\partial x^2} + \frac{\partial^2 v}{\partial y^2} + \frac{\partial^2 v}{\partial z^2}\right) \tag{B-13}$$
$$+ \tfrac{1}{3}\nu\frac{\partial}{\partial y}\left(\frac{\partial u}{\partial x} + \frac{\partial v}{\partial y} + \frac{\partial w}{\partial z}\right)$$

and

$$\frac{\partial w}{\partial t} + u\frac{\partial w}{\partial x} + v\frac{\partial w}{\partial y} + w\frac{\partial w}{\partial z} = f_z - \frac{1}{\rho}\frac{\partial p}{\partial z}$$
$$+ \nu\left(\frac{\partial^2 w}{\partial x^2} + \frac{\partial^2 w}{\partial y^2} + \frac{\partial^2 w}{\partial z^2}\right) \tag{B-14}$$
$$+ \tfrac{1}{3}\nu\frac{\partial}{\partial z}\left(\frac{\partial u}{\partial x} + \frac{\partial v}{\partial y} + \frac{\partial w}{\partial z}\right)$$

The system is completed by a continuity equation which can be developed as in Sec. 3-2.

$$\frac{\partial u}{\partial x} + \frac{\partial v}{\partial y} + \frac{\partial w}{\partial z} = 0 \qquad \text{(B-15)}$$

Thus the quantity in parentheses in the last term of each equation above is zero for incompressible flow.

A few observations about these equations are in order. First, there are three coupled equations. Even for a planar or axisymmetric flow, one must treat two coupled equations. Second, the nonlinear terms are the *inviscid* terms on the left-hand sides. Note that these terms are not simplified by the boundary layer approximations. Third, these equations are rendered *elliptic* by the addition of such terms as $\partial^2 u/\partial x^2$ (and $\partial^2 u/\partial z^2$) in the x momentum equation, Eq. (B-12), while the corresponding equation with the boundary layer approximation is *parabolic*. See Sec. 3-6 for a brief discussion of some of the implications of that distinction.

REFERENCES

ABRAMOVICH, G. N., *The Theory of Turbulent Jets*, MIT Press, Cambridge, Mass., 1960 (English edition).

ACHARYA, M., "Effects of Compressibility on Boundary-Layer Turbulence," AIAA Paper 76-334 (1976).

ANDERSON, E. C. and LEWIS, C. H., "Laminar or Turbulent Boundary-Layer Flows of Perfect Gases or Reacting Gas Mixtures in Chemical Equilibrium," NASA CR-1893 (1971).

ANTONIA, R. A. and BILGER, R. W., "An Experimental Investigation of an Axisymmetric Jet in a Co-flowing Air Stream," *J. Fluid Mech.*, Vol. 61, pp. 805–822 (1973).

ANTONIA, R. A., PRABHU, A., and STEPHENSON, S. F., "Conditionally Sampled Measurements in a Heated Turbulent Jet," *J. Fluid Mech.*, Vol. 72, pp. 455–480 (1975).

BACK, L. H., CUFFEL, R. F., and MESSIER, P. F., "Laminarization of a Turbulent Boundary Layer in Nozzle Flow," *AIAA J.*, Vol. 7, pp. 730–733 (1969).

BECKER, H. A., HOTTEL, H. C., and WILLIAMS, G. C., "The Nozzle Fluid Concentration Field of the Round Turbulent Free Jet," *J. Fluid Mech.*, Vol. 30, pp. 285–303 (1967).

BELOV, V. M., "Experimental Investigation of Heat Transfer in a Turbulent Boundary Layer with a Step-like Change in Thermal Boundary Conditions on the Wall," Thesis (Cand. Sci.), The Bauman Higher Technical College, 1976.

BIRCH, A. D., BROWN, D. R., DODSON, M. G., and THOMAS, J. R., "The Turbulent Concentration Field of a Methane Jet," *J. Fluid Mech.*, Vol. 88, pp. 431–449 (1978).

BLACKWELDER, R. F. and KAPLAN, R. E., "On the Wall Structure of the Turbulent Boundary Layer," *J. Fluid Mech.*, Vol. 76, pp. 89–112 (1976).

BLASIUS, H., "Das Ähnlichkeitsgesetz bei Reibungsvorgangen in Flüssigkeiten," *Forsch. Arb. Ingenieurwes.*, No. 131, Berlin (1913).

BLASIUS, H., "Grenzschichten in Flüssigkeiten mit kleiner Reibung," *Z. Math. Phys.*,

Vol. 56, No. 1, pp. 1–37 (1908). [Available in translation as NACA TM 1256 (1950).]

BLOM, J., "An Experimental Determination of the Turbulent Prandtl Number in a Developing Temperature Boundary Layer," Technische Hogeschool, Eindhoven, 1970.

BLOTTNER, F. G., "Finite Difference Methods of Solution of the Boundary-Layer Equations," *AIAA J.*, Vol. 8, pp. 193–205 (1970).

BOUSSINESQ, J., "Théorie de l'écoulement tourbillant," *Mem. Pres. Acad. Sci., Paris*, Vol. 23, p. 46 (1877).

BRADSHAW, P., FERRISS, D. H., and ATWELL, N. P., "Calculation of Boundary Layer Development Using the Turbulence Energy Equation," *in* (Coles and Hirst) *Computation of Turbulent Boundary Layers—1968 AFOSR-IFP-Stanford Conference*, Stanford University Press, Stanford, Calif., 1969. Original paper published in *J. Fluid Mech.*, Vol. 28, pp. 539–616 (1967).

BRINICH, P. F., "Boundary Layer Transition at Mach 3.12 with and without Single Roughness Element," NACA TN 3267 (1954).

BRINICH, P. F. and DIACONIS, N. S., "Boundary Layer Development and Skin Friction at Mach-Number 3.05," NACA TN 2742 (1952).

BUDDENBERG, J. W. and WILKE, C. R., "Calculation of Gas Mixture Viscosities," *Ind. Eng. Chem.*, Vol. 41, pp. 1345–1347 (1949).

BUSEMANN, A., "Gasdynamik," *Handbuch der Experimentalphysik*, Vol. IV, Pt. 1, Leipzig, 1931.

BUSHNELL, D. and BECKWITH, I., "Calculation of Non-equilibrium Hypersonic Turbulent Boundary Layers and Comparisons with Experimental Data," *AIAA J.*, Vol. 8, pp. 1462–1469 (1970).

CARSLAW, H. S. and JAEGER, J. C., *Conduction of Heat in Solids*, Clarendon Press, Oxford, 1959.

CEBECI, T., "Calculation of Compressible Turbulent Boundary Layers with Heat and Mass Transfer," *AIAA J.*, Vol. 9, pp. 1091–1097 (1971).

CEBECI, T. and MOSINSKIS, G. J., "Prediction of Turbulent Boundary Layers with Mass Addition, Including Highly Accelerating Flows," ASME 70-HT/SpT-19 (1970).

CEBECI, T. and SMITH, A. M. O., "A Finite-difference Solution of the Incompressible Turbulent Boundary Layer Equations by an Eddy Viscosity Concept," *in* (Coles and Hirst) *Computation of Turbulent Boundary Layers—1968 AFOSR-IFP-Stanford Conference*. Stanford University Press, Stanford, Calif., 1969.

CHAPMAN, D. R. and KESTER, R. H., "Measurements of Turbulent Skin Friction on Cylinders in Axial Flow at Subsonic and Supersonic Velocities," *J. Aerosp. Sci.*, Vol. 20, pp. 441–448 (1953).

CHEVRAY, R. and TUTU, N. K., "Intermittency and Preferential Transport of Heat in a Round Jet," *J. Fluid Mech.*, Vol. 88, pp. 133–160 (1978).

CHRISS, D. E., "Experimental Study of Turbulent Mixing of Subsonic Axisymmetric Gas Streams," Arnold Engineering Development Center, AEDC-TR-68-133 (Aug. 1968).

CHRISTIAN, W. J. and KEZIOS, S. P., "Sublimation from Sharp-Edged Cylinders in Axisymmetric Flow, Including Influence of Surface Curvature," *AIChE J.*, Vol. 5, pp. 61–68 (1959).

CLAUSER, F. H., "Turbulent Boundary Layers in Adverse Pressure Gradients," *J. Aerosp. Sci.*, Vol. 21, pp. 91–108 (1954).

CLAUSER, F. H., "The Turbulent Boundary Layer," in *Advances in Applied Mechanics*, Vol. IV, Academic Press, New York, 1956.

COLES, D., "Measurements of Turbulent Friction on a Smooth Flat Plate in Supersonic Flows," *J. Aerosp. Sci.*, Vol. 21, pp. 433–448 (1954).

COLES, D., "The Law of the Wall in Turbulent Shear Flow," in *50 Jahre Grenzschichtforschung*, F. Vieweg & Sohn, Braunschwieg, 1955.

COLES, D., "The Law of the Wake in the Turbulent Boundary Layer," *J. Fluid Mech.*, Vol. 1, pp. 191–226 (1956).

COLES, D., "The Turbulent Boundary Layer in a Compressible Fluid," The Rand Corp., Rep. R-403-PR (1962).

COLES, D. and HIRST, E. A., *Computation of Turbulent Boundary Layers—1968 AFOSR–IFP–Stanford Conference*, Stanford University Press, Stanford, Calif., 1969.

COLLIER, F. S. and SCHETZ, J. A., "Injection into a Turbulent Boundary Layer through Porous Walls with Different Surface Geometries," AIAA Paper No. 83-0295 (Jan. 1983).

CORRSIN, S., and KISTLER, A. L., "The Free-Stream Boundaries of Turbulent Flows," NACA TN 3133 (1954).

CORRSIN, S. and UBEROI, M., "Further Experiments on the Flow and Heat Transfer in a Heated Turbulent Air Jet," NACA TN 1895 (1949).

CROCCO, L., "Sulla transmissione del calore da una lamina piana a un fluido scorrente ad alta velocita," *L'Aerotechnica*, Vol. 12, fasc. 2, pp. 181–197 (Feb. 1932). [Available in translation as NACA TM 690 (1932).]

CROCCO, L., *Lo strato limite laminare nei gas*, Monografie Scientifiche di Aeronautica No. 3, Ministero della Difesa-Aeronautics, Roma, Dec. 1946. [Trans. in North American Aviation Aerophysics Lab. Rep. AL-684 (July 15, 1948).]

DALY, B. J. and HARLOW, F. H., "Transport Equations in Turbulence," *Phys. Fluids*, Vol. 13, p. 2634 (1970).

DEISSLER, R. G., "Analysis of Turbulent Heat Transfer, Mass Transfer and Friction in Smooth Tubes at High Prandtl and Schmidt Numbers," NACA Rep. 1210 (1955).

DHAWAN, S., "Direct Measurements of Skin Friction," NACA Rep. 1121 (1953).

DONALDSON, C. DuP., "A Progress Report on an Attempt to Construct an Invariant Model of Turbulent Shear Flows," *Turbulent Shear Flows*, AGARD CP-93 (1971).

DORODNITSYN, A. A., "Boundary Layer in a Compressible Gas," *Prikl. Math. Mek.*, Vol. 6 (1942).

DRYDEN, H. L., "Airflow in the Boundary Layer near a Plate," NACA Rep. 562 (1936).

DRYDEN, H. L., "Review of Published Data on the Effect of Roughness on Transition from Laminar to Turbulent Flow," *J. Aerosp. Sci.*, Vol. 20, pp. 477–482 (1953).

ECKERT, E. R. G., "Die Berechnung des Wärmeübergangs in der laminaren Grenzschicht," *VDI-Forschungsh.*, No. 416 (1942).

ECKERT, E. R. G., "Engineering Relations for Heat Transfer and Friction in High-Velocity Laminar and Turbulent Flow over Surfaces with Constant Pressure and Temperature," *Trans. ASME*, Vol. 78, pp. 1273–1283 (1956).

ECKERT, E. R. G. and DRAKE, R. M. JR., *Heat and Mass Transfer*, McGraw-Hill, New York, 1959.

ECKERT, E. R. G. and DREWITZ, O., "Der Wärmeubergang an eine mit grosser, Geschwindigkeit längsangeströmte Platte," *Forsch. Ingenieurwes.*, Vol. 11, pp. 116–124 (1940).

ECKERT, E. R. G. and LIEBLEIN, V., "Berechnung des Stoff überganges an einer ebenen, längsangeströmten ober fläche bei grossem Teildruckgefälle," *Forsch. Geb. Ingenieurwes.*, Vol. 16, pp. 33–42 (1949).

ECKERT, E. R. G., HARTNETT, J. P., and BIRKEBAK, R., "Simplified Equations for Calculating Local and Total Heat Flux to Non-isothermal Surface," *J. Aerosp. Sci.*, Vol. 24, pp. 549–551 (1957).

ECKERT, E. R. G., SCHNEIDER, P. J., HAYDAY, A. A., and LARSON, R. M., "Mass-Transfer Cooling of a Laminar Boundary by Injection of a Light-Weight Foreign Gas," *Jet Propul.*, Vol. 3, pp. 34–39 (1958).

EGGERS, J. M., "Turbulent Mixing of Coaxial Compressible Hydrogen-Air Jets," NASA TN D-6487 (1971).

EVERITT, K. W. and ROBINS, A. G., "The Development and Structure of Turbulent Plane Jets," *J. Fluid Mech.*, pp. 563–583 (1978).

FAGE, A. and FALKNER, V. M., "Relation between Heat Transfer and Surface Friction for Laminar Flow," ARC R&M, No. 1408 (1931).

FALKNER, V. M., and SKAN, S. W., "Some Approximate Solutions of the Boundary Layer Equations," ARC R&M, No. 1314 (1930).

FEINDT, E. G., "Untersuchungen über die Abhängigkeit des Umschlages Laminar-Turbulent von der Oberflächenraughigkeit und der Druckverteilung," Diss. Braunschweig, 1956; *Jb. 1956 Schiffbautechn. Ges.*, Vol. 50, pp. 180–203 (1957).

FERRI, A., LIBBY, P. A., and ZAKKAY, V., "Theoretical and Experimental Investigation of Supersonic Combustion," *3rd ICAS Conf.*, Stockholm, 1962.

FLÜGGE-LOTZ, I. and BLOTTNER, F. G., "Computation of the Compressible Laminar Boundary-Layer Flow Including Displacement Thickness Interaction Using Finite Difference Methods," Stanford Univ. Tech. Rep. 1313 (1962).

FORSTALL, W., JR. and SHAPIRO, A. H., "Momentum and Mass Transfer in Coaxial Gas Jets," *J. Appl. Mech.*, Vol. 72, pp. 339–408 (1950).

FULACHIER, L., "Contribution à l'étude des analogies des champs dynamique et thermique dans une couche limite turbulent, effet de l'aspiration," Thèse (Doc. Sci.), Phys. Univ. Provence, Marseille, 1972.

GIBSON, M. M., "Spectra of Turbulence in a Round Jet," *J. Fluid Mech.*, Vol. 15, pp. 161–173 (1963).

GÖRTLER, H., "Berechnung von Aufgaben der freien Turbulenz auf Grund eines neuen Näherungsansatzes," *Z. Angew. Math. Mech.*, Vol. 22, pp. 244–254 (1942).

HALL, A. A. and HISLOP, G. S., "Experiments on the Transition of the Laminar Boundary Layer on a Flat Plate," ARC R&M, No. 1843 (1938).

HAMA, F. R., "Boundary Layer Characteristics for Smooth and Rough Surfaces," *Trans. Soc. Nav. Arch. Mar. Eng.*, Vol. 62, pp. 333–358 (1954).

HANJALIC, K. and LAUNDER, B. E., "A Reynolds Stress Model of Turbulence and Its Application to Asymmetric Shear Flows," *J. Fluid Mech.*, Vol. 52, p. 609 (1972).

HARSHA, P. T., "Prediction of Free Turbulent Mixing Using a Turbulent Kinetic Energy Method," *Free Turbulent Shear Flows*, NASA SP-321 (1971a).

HARSHA, P. T., "Free Turbulent Mixing: A Critical Evaluation of Theory and Experiment," *Turbulent Shear Flows*, AGARD CP-93 (1971b).

HARTNETT, J. P. and ECKERT, E. R. G., "Mass Transfer Cooling in a Laminar Boundary Layer with Constant Fluid Properties," *Trans. ASME*, Vol. 79, pp. 247–254 (1957).

HARTREE, D. R., "On an Equation Occurring in Falkner and Skan's Approximate Treatment of the Equations of the Boundary Layer," *Proc. Camb. Philos. Soc.*, Vol. 33, p. 223 (1937).

HERRING, H. J. and MELLOR, G. L., "A Method of Calculating Compressible Turbulent Boundary Layers," NASA CR-114 (1968).

HILL, F. K., "Boundary-Layer Measurements in Hypersonic Flow," *J. Aerosp. Sci.*, Vol. 23, pp. 35–42 (1956).

HINZE, J. L., *Turbulence*, McGraw-Hill, New York, 1959.

HOPKINS, E. J., KEENER, E. R., POLEK, T. E., and DWYER, H. A., "Hypersonic Turbulent Skin-Friction and Boundary-Layer Profiles on Nonadiabatic Flat Plates," *AIAA J.*, Vol. 10, pp. 40–48 (1972).

HOSSAIN, M. S., "Mathematische Modellierung von turbulenten Auftriebsströmungen," Ph.D. thesis, University of Karlsruhe, 1979.

HOWARTH, L., "Concerning the Effect of Compressibility on Laminar Boundary Layers and Their Separation," *Proc. R. Soc. Lond.*, Vol. A194, p. 16 (1948).

HUFFMAN, G. D., ZIMMERMAN, D. R., and BENNETT, W. A., "The Effect of Free-Stream Turbulence Level on Turbulent Boundary Layer Behavior," in *Boundary Layer Effects in Turbomachines*, AGARDograph 164 (Dec. 1972).

JACK, J. R. and DIACONIS, W. S., "Variation of Boundary-Layer Transition with Heat Transfer at Mach Number 3.12," NACA TN 3562 (1955).

JAFFE, N. A., OKAMURA, T. T., and SMITH, A. M. O., "Determination of Spatial Amplification Factors and Their Application to Predicting Transition," *AIAA J.*, Vol. 8, pp. 301–308 (1970).

JONES, W. P. and LAUNDER, B. E., "The Prediction of Laminarization with a Two-Equation Model of Turbulence," *Int. J. Heat Mass Trans.*, Vol. 15, pp. 301–314 (1972).

JULIEN, H., "The Turbulent Boundary Layer on a Porous Plate: Experimental Study of the Effects of a Favorable Pressure Gradient," Ph.D. thesis, Stanford University, Apr. 1969.

KADER, B. A., "Temperature and Concentration Profiles in Fully Turbulent Boundary Layers," *Int. J. Heat Mass Trans.*, Vol. 24, pp. 1541–1544 (1981).

KARPLUS, W. J., "An Electric Circuit Theory Approach to Finite Difference Stability," *Trans. AIEE*, Vol. 77, pp. 210–213 (1958).

KISTLER, A. L., "Fluctuation Measurements in a Supersonic Turbulent Boundary Layer," *Phys. Fluids*, Vol. 2, pp. 290–296 (1959).

KLEBANOFF, P. S., "Characteristics of Turbulence in a Boundary Layer with Zero Pressure Gradient," NACA Rept. 1247 (1955).

KLEBANOFF, P. S. and DIEHL, F. W., "Some Features of Artificially Thickened Fully

Developed Turbulent Boundary Layers with Zero Pressure Gradient," NACA TN 2475 (1951).

KONG, F. and SCHETZ, J. A., "Turbulent Boundary Layer over Solid and Porous Surfaces with Small Roughness," AIAA Paper 81-0418 (Jan. 1981).

KONG, F. and SCHETZ, J. A., "Turbulent Boundary Layer over Porous Surfaces with Different Surface Geometries," AIAA Paper 82-0030 (Jan. 1982).

KORKEGI, R. H., "Transition Studies and Skin-Friction Measurements on an Insulated Flat Plate at a Mach-Number of 5.8," *J. Aerosp. Sci.*, Vol. 23, pp. 97–102 (1956).

LANDIS, F. and SHAPIRO, A. H., "The Turbulent Mixing of Co-axial Gas Jets," Heat Transfer and Fluid Mechanics Inst., Preprints and Papers, Stanford University Press, Stanford, Calif., 1951.

LAUFER, J., "Investigations of Turbulent Flow in a Two-Dimensional Channel," NACA TR 1053 (1951).

LAUFER, J., "The Structure of Turbulence in Fully Developed Pipe Flow," NACA TR 1174 (1954).

LAUNDER, B. E. and SPALDING, D. B., *Mathematical Models of Turbulence*, Academic Press, New York, 1972.

LAUNDER, B., MORSE, A., RODI, W., and SPALDING, D. B., "Prediction of Free Shear Flows—A Comparison of the Performance of Six Turbulence Models," *Free Turbulent Shear Flows*, NASA SP-321 (1971).

LEE, R. E., YANTA, W. J., and LEONAS, A. C., "Velocity Profile, Skin-Friction Balance and Heat-Transfer Measurements of the Turbulent Boundary Layer at Mach 5 and Zero-Pressure Gradient," Nav. Ord. Lab. Rep. TR-69-106 (1969).

LEES, L., "Laminar Heat Transfer over Blunt-Nosed Bodies at Hypersonic Flight Speeds," *Jet Propul.*, Vol. 26, pp. 259–268 (1956).

LEVY, S., "Heat Transfer to Constant-Property Laminar Boundary-Layer Flows with Power-Function Free-Stream Velocity and Wall-Temperature Variation," *J. Aerosp. Sci.*, Vol. 19, p. 341 (1952).

LIN, C. C., "On the Stability of Two-Dimensional Parallel Flows," *Q. Appl. Math.*, Vol. 3, pp. 277–301 (1945).

LOBB, R. K., WINKLER, E. M., and PERSH, J., "Experimental Investigation of Turbulent Boundary Layers in Hypersonic Flow," NAVORD Rep. 3880 (1955).

LOW, G. M., "Cooling Requirement for Stability of Laminar Boundary Layer with Small Pressure Gradient at Supersonic Speeds," NACA TN 3103 (1954); see also *J. Aerosp. Sci.*, Vol. 22, pp. 329–336 (1955).

LUDWIEG, H. and TILLMANN, W., "Investigation of the Wall Shearing Stress in Turbulent Boundary Layers," NACA TM 1285 (1950).

MACK, L. M., "An Experimental Investigation of the Temperature Recovery-Factor," Jet Propulsion Lab., Rep. 20-80, California Institute of Technology, Pasadena (1954).

MAISE, G. and McDONALD, H., "Mixing Length and Kinematic Eddy Viscosity in a Compressible Boundary Layer," *AIAA J.*, Vol. 6, pp. 73–80 (1968).

MICHEL, R., "Étude de la transition sur les profiles d'aile; éstablissement d'un critère de détermination de point de transition et calcul de la trainée de profile incompressible," ONERA Rep. 1/1578A (1952).

MICKLEY, H. S., ROSS, R. C., SQUYERS, A. L., and STEWART, W. E., "Heat, Mass and

Momentum Transfer for Flow over a Flat Plate with Blowing and Suction," NACA TN 3208 (1954).

MINER, E. W., ANDERSON, E. C., and LEWIS, C. H., "A Computer Program for Two-Dimensional and Axisymmetric Nonreacting Perfect Gas and Equilibrium Chemically Reacting Laminar, Transitional and-or Turbulent Boundary Layer Flows," NASA CR-132601 (1975).

MONAGHAN, R. J. and COOKE, J. R., "The Measurement of Heat Transfer and Skin Friction at Supersonic Speeds—Part IV Test on a Flat Plate at $M = 2.82$," RAE Tech. Note Aero. 2171 (1952).

MOODY, L. F., "Friction Factors for Pipe Flow," Trans. ASME, Vol. 66, pp. 671 (1944).

MOSES, H. L., "A Strip-Integral Method for Predicting the Behavior of Turbulent Boundary Layers," in (Coles and Hirst) Computation of Turbulent Boundary Layers—1968 AFOSR-IFP-Stanford Conference, Stanford University Press, Stanford, Calif., 1969.

NAVIER, M., "Mémoire sur les lois du mouvement des fluides," Mem. Acad. Sci., Vol. 6, pp. 389–416 (1823).

NEWMAN, B. G., "Some Contributions to the Study of the Turbulent Boundary Layer near Separation," Australia Dept. Supply Rep. ACA-53 (1951).

NG, K. H. and SPALDING, D. B., "Predictions of Two-Dimensional Boundary Layers on Smooth Walls with a Two-Equation Model of Turbulence," Rept. cw/16, Imperial College, London (1970).

NG, K. H., PATANKAR, S. V., and SPALDING, D. B., "The Hydrodynamic Boundary Layer on a Smooth Wall Calculated by a Finite-Difference Method," in (Coles and Hirst) Computation of Turbulent Boundary Layers—1968 AFSOR-IFP-Stanford Conference, Stanford University Press, Stanford, Calif., 1969.

NIKURADSE, J., "Gesetzmässigkeit der turbulenten Strömung in glatten Rohren," VDI-Forschungsh., No. 356 (1932).

NIKURADSE, J., Laminare Reibungsschichten an der längsangeströmten Platte, Monograph, Zentrale für Wissenschaft, Berichtwesen, Berlin, 1942.

PARR, W., "Laminar Boundary Layer Calculations by Finite Differences," Nav. Ord. Lab. Rep. TR-63-261 (1963).

PATANKAR, S. V. and SPALDING, D. B., Heat and Mass Transfer in Boundary Layers, Intertext Books, London, 1967.

PIERCY, N. A. V. and PRESTON, J. H., "A Simple Solution of the Flat Plate Problem of Skin Friction and Heat Transfer," Philos. Mag. (7), Vol. 21, pp. 995–1005 (1936).

POHLHAUSEN, K., "Der Wärmeaustausch zwischen festen Körpern und Flüssigkeiten mit kleiner Reibung und kleiner Wärmeleitung," Z. Angew. Math. Mech., Vol. 1, p. 115 (1921a).

POHLHAUSEN, K., "Zur näherungsweisen Integration der Differentialgleichungen der laminaren Reibungsschicht, " Z. Angew. Math. Mech., Vol. 1, pp. 252–268 (1921b).

POTTER, J. L. and WHITFIELD, J. D., "Effects of Slight Nose Bluntness and Roughness on Boundary-Layer Transition in Supersonic Flows," J. Fluid Mech., Vol. 12, pp. 501–535 (1962).

PRANDTL, L., "Über Flüssigkeitsbewegung bei sehr kleiner Reibung," Proc. 3rd Int. Math. Congr., Heidelberg, 1904; see also L. Prandtl, "Gesammelte Abhandlungen zur angewandten Mechanik," *Hydro- und Aerodynamik* (Collected Works), W. Tollmien, H. Schlichting, and H. Görtler (eds.), Vol. 2, Springer-Verlag, Berlin, 1961.

PRANDTL, L., "Über die ausgebildete Turbulenz," *Z. Angew. Math. Mech.*, Vol. 5, pp. 136–139 (1925).

PRANDTL, L., *Ergeb. AVA Goettingen*, Ser. III, pp. 1–5 (1927).

PRANDTL, L., "The Mechanics of Viscous Fluids," in W. F. Durand, *Aerodynamic Theory, III*, 1935; see also summary by L. Prandtl, "Neuere Ergebnisse der Turbulenzforschung," *VDI Z.*, Vol. 77, pp. 105–114 (1933); see also *Collected Works*, Vol. 2, pp. 819–845.

PRANDTL, L., "Bemerkungen zur Theorie der freien Turbulenz," *Z. Angew. Math. Mech.*, Vol. 22, pp. 241–243 (1942).

PRANDTL, L., "Über eine neues Formelsystem für die ausgebildete Turbulenz," *Nachr. Akad. Wiss., Goettingen, Math. Phys. Ke.*, pp. 6–19 (1945).

PRANDTL, L. and REICHARDT, H., "Einfluss von Wärmeschichtung auf die Eigenschaften einer turbulenten Stromung," *Dtsch. Forsch.*, No. 21, pp. 110–121 (1934); see also *Collected Works*, Vol. 2, pp. 846–854.

REICHARDT, H., "Gesetzmassigkeitender freien Turbulenz," *VDI-Forschungsh.*, No. 414 (1942); 2nd ed. (1951).

REICHARDT, H., "Vollstandige Darstellung der turbulenten Geschwindigkeitsverteilung in glatten Leitungen," *Z. Angew. Math. Mech.*, Vol. 31, pp. 208–219 (1951).

REYNOLDS, O., "An Experimental Investigation of the Circumstances Which Determine Whether the Motion of Water Shall Be Direct or Sinuous, and of the Law of Resistance in Parallel Channels," *Philos. Trans. R. Soc. Lond.*, Vol. 174, pp. 935–982 (1883).

REYNOLDS, W. C., KAYS, W. M., and KLINE, S. J., "Heat Transfer in a Turbulent Incompressible Boundary Layer," NASA Memo 12-4-58W (1958).

RODI, W., "The Prediction of Free Turbulent Boundary Layers by Use of a Two-Equation Model of Turbulence," Ph.D. thesis, University of London, 1972.

RODI, W., "A Review of Experimental Data of Uniform Density Free Turbulent Boundary Layers," in *Studies in Convection*, Vol. 1, B. E. Launder (ed.), Academic Press, London, 1975.

RUBESIN, M., M. S. thesis, University of California, Berkeley, 1949.

RUBESIN, M. W., CRISALLI, A. J., LANFRANCO, M. J., and ACHARYA, M., "A Critical Evaluation of Invariant Second-Order Closure Models for Subsonic Boundary Layers," *Proceedings of the Symposium on Turbulent Shear Flows* (Pennsylvania State University, 1977), Spring-Verlag, Berlin, 1979.

SCHETZ, J. A., "Supersonic Diffusion Flames," *Supersonic Flow, Chemical Processes and Radiative Transfer*, Pergamon Press, London, 1964.

SCHETZ, J. A., "Turbulent Mixing of a Jet in a Coflowing Stream," *AIAA J.*, Vol. 6, pp. 2008–2010 (1968).

SCHETZ, J. A., "Analysis of the Mixing and Combustion of Gaseous and Particle Laden Jets in an Airstream," AIAA Paper 69-33 (1969).

SCHETZ, J. A., "Some Studies of the Turbulent Wake Problem," *Astronaut. Acta*, Vol. 16, pp. 107–117 (1971a).

SCHETZ, J. A., "Free Turbulent Mixing in a Co-flowing Stream," *Free Turbulent Shear Flows*, NASA SP-321 (1971b).

SCHETZ, J. A., *Injection and Mixing in Turbulent Flow*, AIAA, New York, 1980.

SCHETZ, J. A. and FAVIN, S., "Numerical Calculation of Turbulent Boundary Layer with Suction or Injection and Binary Diffusion," *Astronaut. Acta*, Vol. 16, pp. 339–352 (1971).

SCHETZ, J. A., BILLIG, F. S., and FAVIN, S., "Flowfield Analysis of a Scramjet Combustor with a Coaxial Fuel Jet," *AIAA J.*, Vol. 20, pp. 1268–1274 (1982).

SCHETZ, J. A. and NERNEY, B., "The Turbulent Boundary Layer with Injection and Surface Roughness," *AIAA J.*, Vol. 15, pp. 1288–1294 (1977).

SCHLICHTING, H., "Über das ebene Windschattenproblem," Diss., Göttingen, 1930; *Ing.-Arch.* Vol. 1, pp. 533–571 (1930).

SCHLICHTING, H., "Turbulenz bei Wärmeschichtung," *Z. Angew. Math. Mech.*, Vol. 15, pp. 313–338 (1935); also *Proc. 4th Int. Congr. Appl. Mech.*, Cambridge, 1935, p. 245.

SCHLICHTING, H., *Boundary Layer Theory*, McGraw-Hill, New York, 1942; 6th ed., 1968.

SCHOENHERR, K. E., "Resistance of Flat Plates Moving through a Fluid," *Trans. Soc. Nav. Arch. Mar. Eng.*, Vol. 40, pp. 279–313 (1932).

SCHUBAUER, G. B., "Turbulent Process as Observed in Boundary Layer and Pipe," *J. Appl. Phys.*, Vol. 25, pp. 188–196 (1954).

SCHUBAUER, G. B. and KLEBANOFF, P. S., "Investigation of Separation of the Turbulent Boundary Layer," NACA TN 2133 (1950).

SCHUBAUER, G. B. and SKRAMSTAD, H. K., "Laminar Boundary-Layer Oscillations and Stability of Laminar Flow," *J. Aerosp. Sci.*, Vol. 14, pp. 69–78 (1947).

SCHUBAUER, G. B. and TCHEN, C. M., *Turbulent Flow*, Princeton University Press, Princeton, N. J., 1961.

SCHULTZ-GRUNOW, F., "A New Resistance Law for Smooth Plates," *Luftfahrt-Forsch.*, Vol. 17, pp. 239–246 (1940).

SCOTT, C. J., ECKERT, E. R. G., JONSSON, V. K., and YANG, Ji-Wu, "Measurements of Velocity and Concentration Profiles for Helium Injection into a Turbulent Boundary Layer Flowing over an Axial Circular Cylinder," University of Minnesota HTL-TR-55 (Feb. 1964).

SEIFF, A., "Examination of the Existing Data on the Heat Transfer of Turbulent Boundary Layers at Supersonic Speeds from the Point of View of Reynolds Analogy," NACA TN 3284 (1954).

SIEDER, E. N. and TATE, G. E., "Heat Transfer and Pressure Drop of Liquid in Tubes," *Ind. Eng. Chem.*, Vol. 28, pp. 1429–1435 (1936).

SIMPSON, R. L., "The Turbulent Boundary Layer on a Porous Wall," Ph.D. thesis, Stanford University, 1968.

SIVASEGARAM, S. and WHITELAW, J. H., "The Prediction of Turbulent Supersonic Two-Dimensional Boundary Layer Flows," *Aeronaut. Q.*, pp. 274–294 (1971).

SMITH, A. G. and SPALDING, D. B., "Heat Transfer in a Laminar Boundary Layer with Constant Properties and Constant Wall Temperature," *J. R. Aeron. Soc.*, Vol. 62, pp. 60–64 (1958).

SMITH, A. M. O., JAFFE, N. A., and LIND, R. C., "Study of a General Method of Solution to the Incompressible Boundary Layer Equations," Douglas Aircraft Div., Rep. LB52949 (1965).

SQUIRE, H. B., "On the Stability for Three-Dimensional Disturbances of Viscous Fluid Flow between Parallel Walls," *Proc. R. Soc. Lond.*, Vol. A142, pp. 621–628 (1933).

STEVENSON, T. N., "A Law of the Wall for Turbulent Boundary Layers with Suction or Injection," Cranfield Coll. Aero. Rep. 166 (1963).

STOKES, G. G., "On the Theories of Internal Friction of Fluids in Motion," *Trans. Camb. Philos. Soc.*, Vol. 8, pp. 287–305 (1845).

STOKES, G. G., "On the Effect of the Internal Friction of Fluids on the Motion of Pendulums," *Trans. Camb. Philos. Soc.*, Vol. 9 (1851).

STUART, J. T., Chap. 9 in *Laminar Boundary Layers*, L. Rosenhead (ed.), Clarendon Press, Oxford, 1963.

STÜPER, J., "Untersuchung von Reibungsschichten am fliegenden Flugzeug," NACA TM 751 (1934).

SUTHERLAND, W., "The Viscosity of Gases and Molecular Force," *Philos. Mag.*, Ser. 5, pp. 507–531 (1893).

SVEHLA, R. A. and McBRIDE, B. J., "FORTRAN IV Computer Program for Calculation of Thermodynamic and Transport Properties of Complex Chemical Systems," NASA TN D-7056 (1973).

TAYLOR, G. I., "The Transport of Vorticity and Heat through Fluids in Turbulent Motion," Appendix by A. Fage and V. M. Faulkner, *Proc. R. Soc. Lond.*, Vol. A135, pp. 685–705 (1932); see also *Philos. Trans.*, Vol. A215, pp. 1–26 (1915).

THWAITES, B., "Approximate Calculation of the Laminar Boundary Layer," *Aeronaut. Q.*, Vol. 1, pp. 245–280 (1949).

TING, I. and LIBBY, P. A., "Remarks on the Eddy Viscosity in Compressible Mixing Flows," *J. Aerosp. Sci.*, Vol. 27, pp. 797–798 (1960).

TOLLMIEN, W., "Berechnung turbulenter Ausbreitungsvorgänge," *Z. Angew. Math. Mech.*, Vol. 6, pp. 468–478 (1926).

TOLLMIEN, W., "Über die Enstehung der Turbulenz," *Nachr. Ges. Wiss. Goettingen*, pp. 21–44 (1929). [Translated as "The Production of Turbulence," NACA TM 609 (1931).]

TOULOUKIAN, Y. S. and HO, C. Y., "*Thermophysical Properties of Matter—The TPRC Data Series*, Y. S. Touloukian and C. Y. Ho (eds.), IFI/Plenum Data Co., New York, Vol. 3, 1970, p. 120; Vol. 6, 1970, p. 102; and Vol. 11, 1975, p. 94.

VAN DRIEST, E. R., "Turbulent Boundary Layer in Compressible Fluids," *J. Aerosp. Sci.*, Vol. 18, pp. 145–160 (1951).

VAN DRIEST, E. R., "Investigation of Laminar Boundary Layer Compressible Fluids Using the Crocco Method," NACA TN 2597 (1952).

VAN DRIEST, E. R., "On Turbulent Flow near a Wall," *J. Aerosp. Sci.*, Vol. 23, pp. 1007–1012 (1956a).

VAN DRIEST, E. R., "The Problem of Aerodynamic Heating," *Aeronaut. Eng. Rev.*, Vol. 15, pp. 26–41 (1956b).

VAN DRIEST, E. R. and BOISON, J. C., "Experiments on Boundary Layer Transition at Supersonic Speeds," *J. Aerosp. Sci.*, Vol. 24, pp. 885–899 (1957).

VAN DRIEST, E. R. and BLUMER, C. B., "Boundary Layer Transition, Free Stream

Turbulence, and Pressure Gradient Effects," *AIAA J.*, Vol. 1, pp. 1303–1306 (1963).

VON KÁRMÁN, TH., "Mechanische Ähnlichkeit und Turbulenz," *Nachr. Ges. Wiss. Goettingen, Math. Phys. Kl.*, p. 58 (1930), and *Proc. 3rd Int. Congr. Appl. Mech.*, Stockholm, Pt. I, 1930, p. 85; NACA TM 611 (1931).

VON KÁRMÁN, TH., "The Analogy between Fluid Friction and Heat Transfer," *Trans. ASME*, Vol. 61, pp. 705–710 (1939).

VON NEUMANN, J.: see G. G. O'Brien, M. S. Hyman, and S. Kaplan, *J. Math. Phys.*, Vol. 29, pp. 223–51 (1952).

WALTRUP, P. J. and SCHETZ, J. A., "Supersonic Turbulent Boundary Layer Subjected to Adverse Pressure Gradients," *AIAA J.*, Vol. 11, pp. 50–57 (1973).

WALZ, A., "Ein neuer Ansatz für das Geschwindigkeitsprofil der laminaren Reibungsschicht," *Ber. Lilienthal-Ges. Luftfahrtf.* No. 141, pp. 8–12 (1941).

WAZZAN, A. R., OKAMURA, T. T., and SMITH, A. M. O., "Spatial and Temporal Stability Charts for the Falkner–Skan Boundary-Layer Profiles," McDonnell-Douglas Rep. DAC-67086 (1968).

WEINSTEIN, A. S., OSTERLE, J. F., and FORSTALL, W., "Momentum Diffusion from a Slot Jet into a Moving Secondary," *J. Appl. Mech.*, pp. 437–443 (1956).

WELLS, C. S., JR., "Effects of Freestream Turbulence on Boundary-Layer Transition," *AIAA J.*, Vol. 5, pp. 172–174 (1967).

WERLÉ, H., private communication, 1982.

WHITE, F. M., *Viscous Fluid Flow*, McGraw-Hill, New York, 1974.

WILSON, R. E., "Turbulent Boundary Layer Characteristics at Supersonic Speeds—Theory and Experiment," *J. Aerosp. Sci.*, Vol. 17, pp. 585–594 (1950).

WILSON, R. E., Secs. 13 and 14, *Handbook of Supersonic Aerodynamics*, NAVORD Rep. 1488, Vol. 5, U.S. Government Printing Office, Washington, D. C., 1966.

WU, J. C., "On the Finite Difference Solution of Laminar Boundary Layer Problems," *Proceedings of the 1961 Heat Transfer and Fluid Mechanics Institute*, Stanford University Press, Stanford, Calif., June 1961.

WYGNANSKI, I. and FIEDLER, H. E., "Some Measurements in the Self-Preserving Jet," *J. Fluid Mech.*, Vol. 38, pp. 577–612 (1969).

ZHUKAUSKAS, A. and SLANCHAUSKAS, A., *Heat Transfer in a Turbulent Liquid Flow*, Mintis, Vilnius, 1973.

INDEX